PRINCIPIA ARITHMETICA REVISION

P. Boyes

Published by
BOYES
HULL YORKSHIRE

British Library Cataloguing-in-Publication Data.
A catalogue record for this book is available from the British Library.

Arthur H. Stockwell Ltd., bear no responsibility for the accuracy of events recorded in this book.

By the same author:
Principia Arithmetica or Not Proven — Part One
Principia Arithmetica or Not Proven — Part Two

ISBN 0-7223-3746-9
ISBN 978-0-7223-3746-2
Printed in Great Britain by
Arthur H. Stockwell Ltd.
Torrs Park Ilfracombe
Devon

CONTENTS

INTRODUCTION

Professors Hoyle and Gold, in their theory of space and matter and how the universe was made and expanded, calculated that the amount necessary to maintain it in its expanding state was a creation between the galaxies of one million million million tons of protons per second. We can quite believe that this is so. It is called a steady-state theory, and though we don't believe a great deal of it we do believe that their statement of atomic creation was correct.

In the big-bang theory, which a lot of scientists accept, there is no explanation of where the matter involved came from, or where it originated. Although Hoyle and Gold could not explain this, we think we have some idea. We think, like Professors Hoyle and Gold, that the universe is self-perpetuating. Atoms are being created in it between the galaxies — these being generated by positive forces from them coming together and forming particles which by multiplication create protons. This fact, we believe, is not so remarkable as an acorn developing into a great oak.

We like to think that we have made several things understandable to the ordinary man in the street — much better in this respect than Einstein's, which are very difficult for ordinary people to comprehend.

The first part of these notes should be read very slowly and carefully, in order to grasp some of their implications and how these fit together, even though only decimals and arithmetic are employed.

The subject of atomic energy is all the rage nowadays, but we think it has a simple side to it. Protons — the smallest atoms — are made up of one nucleus and an encircling electron. (The former is positively charged, the latter is negatively charged.) Two of these make up deuterium, one having reversed its positive field charges to negative ones. It's feasible to put two of these deuterium atoms together, these forming then helium, or what are termed alpha particles. These atoms then are inert or negative in their character — the negative particles or neutrons having absorbed the positive fields of the protons. These alphas are, we believe, square in their structure, with the result that larger atoms, which they build up, have a cuboid form, but while this is happening their positive nuclei are being multiplied. This goes on until heavy atoms become unstable. This particularly applies to uranium 235, an atom which has three positive-absorbing neutrons missing so that it has a tendency to split, producing great energy. The same is true of plutonium, an artificial element, which, just like uranium 235, has three neutrons missing, and so fission is liable to occur when chain reactions are induced in it.

All great masses already formed (galaxies) put out positive and negative emanations which we imagine coming together are multiplied, this having several effects. We believe these are atomic. Now, atoms, we know, are composed of nuclei and attendant particles rotating around them.

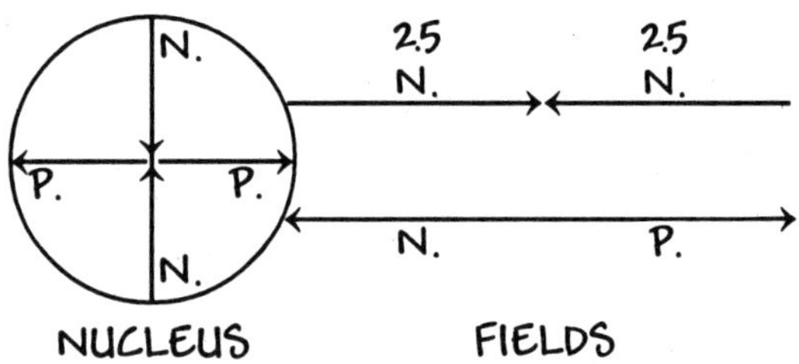

If we look at this figure, we can see all the forces involved in the smallest atom — a proton. Two negatives hold its nucleus together, and these, at the same time, exerting pressure at its centre, cause it to spin rapidly and throw out a centrifugal force, most of which is concentrated on its equator. The fields mirror these internal forces. First we must deal with the negative one. For simplicity, we must assume that its two inwardly directed components are each 2.5 atomic units in strength, and we must as well imagine that they exist all round the nucleus, taking a ring-like form. This means that there are four of them, each consisting of two 2.5s. Their total strength is 20 atomic units and these are quite capable of combining into a particle — a negative particle — this holding together because of its inward tendency, and travelling away from the atomic nucleus into space. Such particles are small, elastic, and very volatile, and their size, we are told by scientists, is compared to the nucleus as 1840 to 1. These particles, being negative, are attracted to anything of a positive nature, and forming currents can become very powerful and destructive. Also they can absorb positive forces and vibrate, giving off what are called electromagnetic radiations, which can vary from radio waves to gamma rays. These latter are very dangerous.

This particle we have been describing is an electron, of course, and it forms half of the fields of protons, which are the smallest atoms.

What about the other half of these fields — the opposite positive halves? These consist of two divisions — positive and negative. They can be called positive extensions, for they extend from the nucleus to the same extent as the electrons, which are of exactly the same strength.

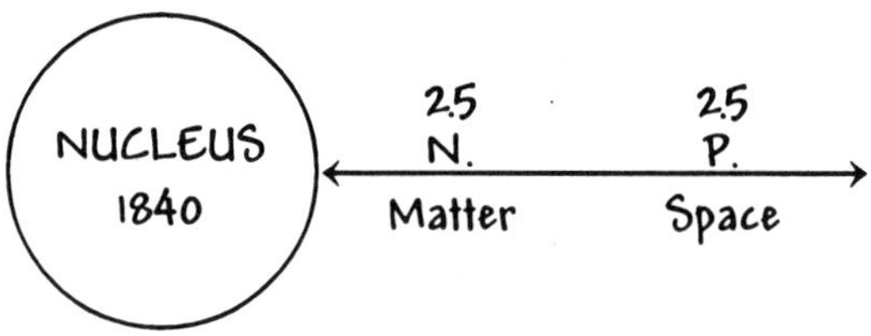

We can observe this in this figure, and can also see how they are divided so as to influence the universe and all that's in it. We believe that the long-range field forces of atoms create space and since there are four of them surrounding atoms, four 2.5 units — atomic units — when multiplied give an indication of space.

$$\begin{array}{r} 2.5 \\ \times 4.0 \\ \hline 10.0 \end{array}$$

Space extends about matter, which we must consider now. As we can observe in the figure, the force comprising matter is close to the nucleus, and this being so it is slightly stronger than that of space — one unit more — which the nucleus, being positive, gives to it. This addition makes its combined force eleven. Now, matter as we know it exists in space, so how can we show that the atomic unit eleven has something to do with it?

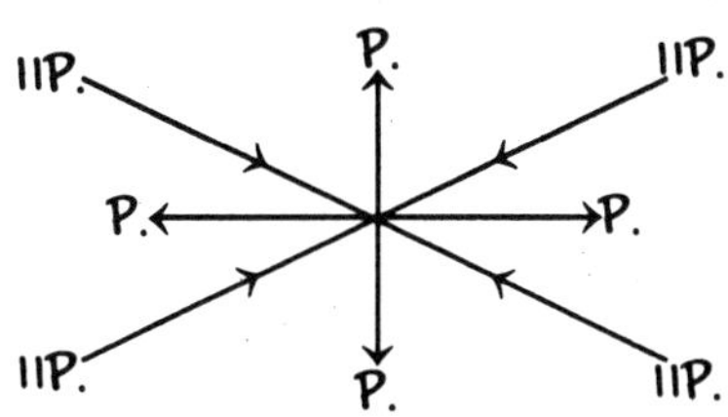

In the preceding illustration it can be seen how four positive forces coming between matter, such as galaxies, can beget by multiplication more forces. Eleven elevens are 121, but 22 times 22 equals 484 and, as we saw before with the electrons, a particle of 20 units resulted from a 2.5 extension. However, we are dealing with positives here, and the inverse square has it that 484 divided by four is 121. Also, in the figure it can be noticed that each extent resulting from the positive collisions is put at the same 121, so that half of the forces we are considering can each be put at 22.

Looking at these resulting four figures, we see that they are exactly like positive magnets, but these, we know, are composed of negative matter, so we can consider them to be just that.

In the diagram, the four 121 units are put very close to the centre of the figure, but in reality we think they will not stay there. They will move away from it and will be replaced by another four, and so on.

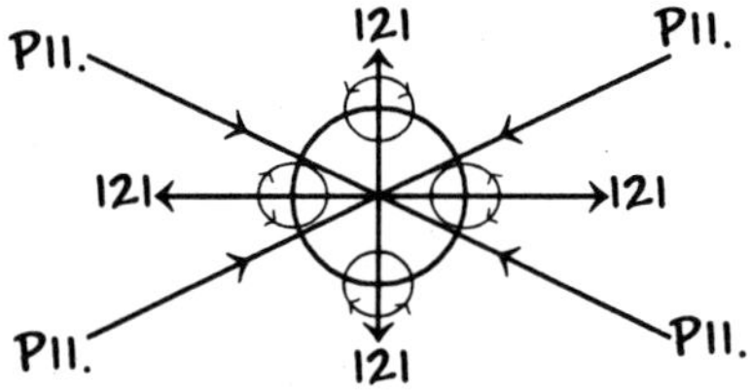

We can call these particles, and they are being generated in space between the galaxies at the rate, according to Professors Hoyle and Gold, of a million million million tons a second. This is the number, they state, which keeps the outward-moving universe in its state of equilibrium.

It is these motes which we believe are half of the components of space.

We should notice, also, here that as the nucleus of atoms is put at 1840 and its attendant particle at 20 units, combining them we have 1860 and it is this figure which we imagine is the strength of each single cycle put out by the electron in response to any influence.

So we have reached a stage when tiny negative particles are released into space. But what is space? We saw previously that each positively charged atom has a long-range positive field and, going on the assumption that this was made up of four positive units, we came to the conclusion that great masses (galaxies) were putting these out. These, we assert, coming together between them were being multiplied 10 times 10. This result amounts to one hundred units. It is obvious then that electromagnetic radiations travelling through space are being multiplied by one hundred, so that 1860 becomes 186,000 — a velocity, which we are assured by scientists is the speed of light and other radiations of electrons.

But what about the created particles? What is their charge? To find this out we divide 121 (11 × 11), this being their size — the 500 their velocity in the cosmos. We divide then 121 by 2:

$$121 \div 2 = 60.5 \qquad 60.5 \times 60.5 = 3660.25 \qquad 3660.25 \div 2 = 1830.12$$

3660.25 is the charge of the 121 particles, but 3660.25, resulting by the fact of its being composed of two opposite halves, contracts by two so that 1830.12 is the figure reached. This, however, is not the finish of the matter. What scientists tell us is that comparing the nucleus's size to the electron 1840 is the correct figure. How is this achieved? But just as the nucleus was divided by two, so we imagine the electron was too. That particle, we know, is negative in its character — that is, its forces are inwardly directed, leading us to the assumption that it contracted when it was being created.

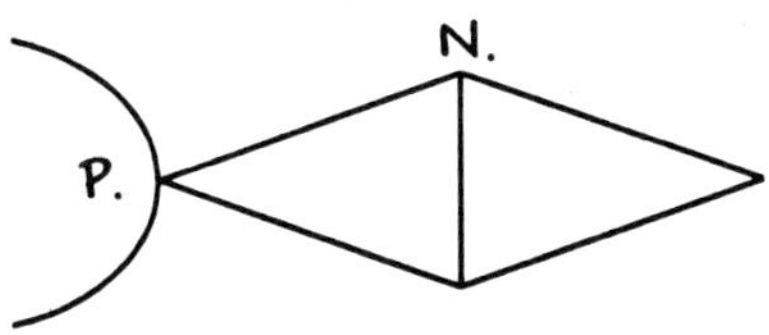

We saw previously that its charges are twenty in their strength, but this would not change with contraction. What would change would be the size of the electron. This would be halved by ten. Its four composite size units are .25, giving the particle an altogether dimension figure of a single unit, so that as the electron became less in size so the nucleus would become larger — by comparison, 1840 to 1.

So we have a picture here of an atom larger in size by 1840 than a particle rotating around it, these created between the composite galaxies.

According to Professors Hoyle and Gold's steady-state theories, such atoms are being created at a vast rate, and it is these attracted into galaxies which constitute interstellar gas — the same gas as they believed was falling continually on the sun, adding a great proportion to its weight over millions of years.

It is our assumption that this gas by combination built up all atoms in it up to the heaviest, and is still doing so up to the present day.

From this account we can observe how space and matter can be created from galaxies.

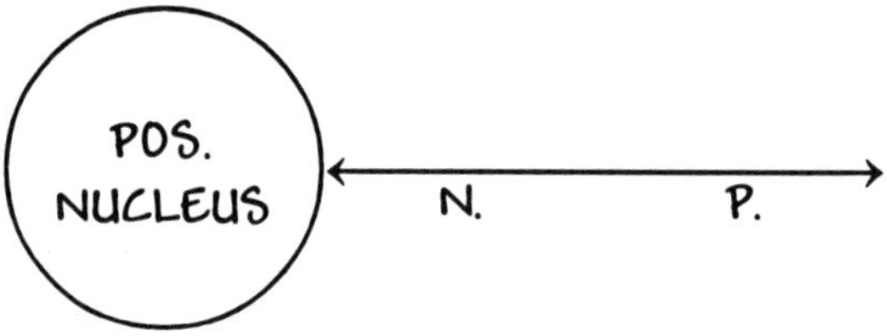

These derived from positively charged atoms. A question might occur here: How do atoms in galaxies give out eleven units, while their external positives creating space add up to so much less? The answer, as is usual in such cases, is simply that close to atoms their fields are separated out, the long-range positives and negatives at a distance are combined so that they add up to eleven.

It is true that in a given area of space and a certain amount of matter these two would weigh exactly the same, both being derived from the same source. So it would appear from the foregoing that the universe is creating itself by a process of multiplication — its components moving away from a centre — and that even space is a substance providing a medium for electromagnetic radiations; and that it itself is a hundred times more rare than electrons, the smallest particle; and that light and radiant heat are exactly similar in space to ripples in ponds or sound in the atmosphere.

We haven't as yet, however, dealt with the undoubted fact of there being three dimensions, and we believe that these are caused in the first place by two opposite forces, positive and negative. We saw that space is positive and this means that it is outward-moving and this begets its contrary force. The situation can be described like this: Suppose we have a piece of string with a knot tied at its centre. The harder the ends of the string are pulled, the tighter the knot becomes. We can say then that this is a negative effect, and if we compare this reaction to space, we can see that this picture emerges:

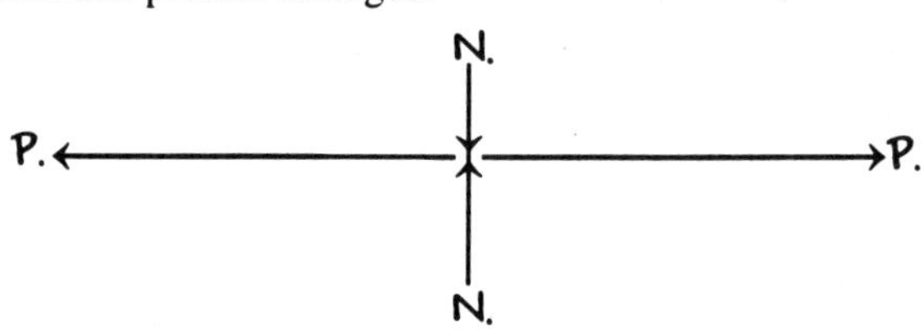

This means that the positive extents are cut in half, and that they are diverted from their straight path by about ninety degrees. If we look at this effect on ordinary matter about us, we can make this diagram:

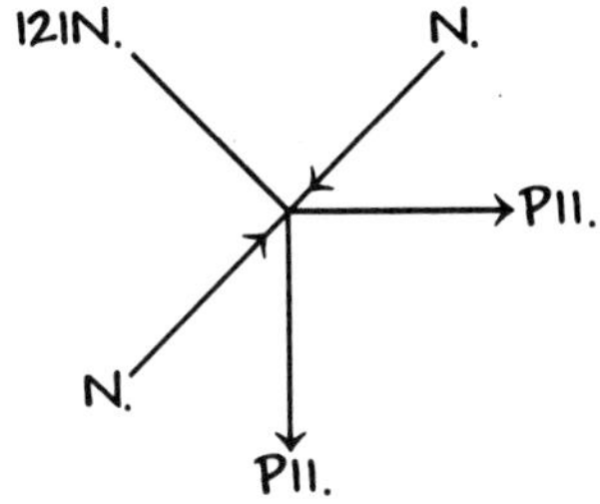

And we can see that an ordinary circle consists of four right-angled figures, so that it is impossible to construct a figure larger in extent than the primary one. We can also make this familiar figure:

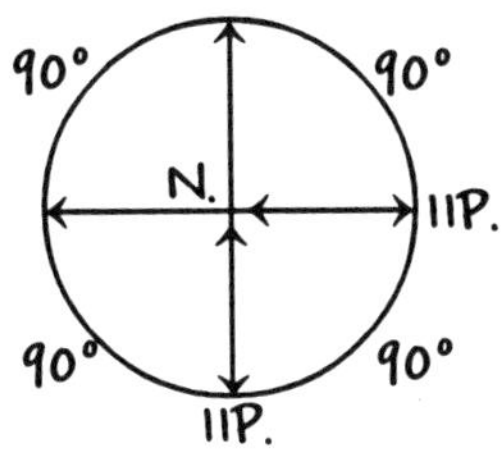

And if we try to improve on it, all we achieve is a multiplication of the forces of the primary one. The negatives produce thickness, and the positives are extensions of it.

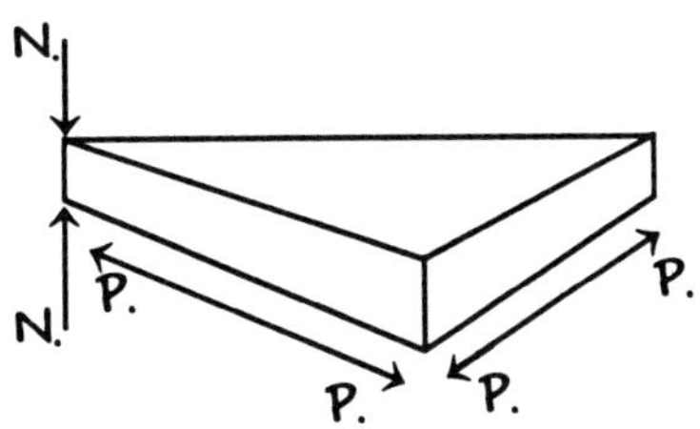

The question may be asked here: Is the negative force going on all round us and existing so as to divert the positives? The answer is, of course, it goes on in the smallest piece of matter to the largest, including the earth and the sun.

In addition to the foregoing, however, we must consider this fact —

that is, if, as pictured, the two positives are at right angles to each other, how are the four cardinals created?

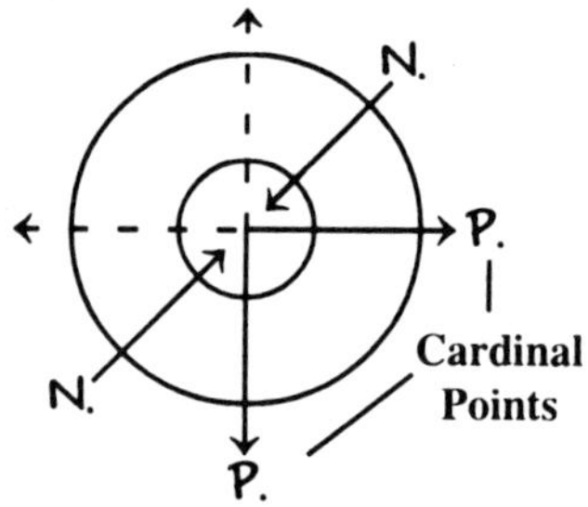

This may seem puzzling, but we think again that the answer to this is simple. It concerns the inverse-square law, which states twice the distance four times the extent. This happens in this instance. Proceeding from a centre, therefore, as in the figure, the distances from it increase by fours, so that instead of there being two cardinals resulting from the action of negatives, there are four, the extent of space increasing by that amount. These are horizontal, just as the negatives are vertical, being inwardly directed and quite oppositely placed regarding the positives. This is made plain in the previous figures when thickness of any matter is indicated.

CHAPTER 1

Part 1

We have reached the stage when atoms have been created in space. We imagine that what happened then was that galaxies of gas were formed and in them pressure being slight, only small elements were produced, these existing in great clouds, mostly ice and smaller elemental matter. During and after this period, these clouds produced masses which eventually evolved into stars, some large, some small. In the case of these latter, some were only a few miles in diameter and were sometimes taken up by the stars as satellites. Such a one is Halley's Comet, the highly eccentric orbit of which seems to show that originally it came from outer space.

When the rotation of the sun is considered, we must always take into account the braking action of its satellites, which is considerable. Normally the sun would rotate quite quickly, as it did when its equatorial disc, which produced its planets, took place. But as soon as they became separate bodies, their brake started to take effect. To try to find the extent of this now, we must use the age-old method of adding them together and then dividing the result by their number, thus obtaining their average.

The planets' main positive force is measured by the length of their diameters. These extents are as follows, starting with Pluto, that planet being the furthest away:

2	Pluto	20	Asteroids
32	Neptune	4	Mars
32	Uranus	8	Earth
72	Saturn	8	Venus
88	Jupiter	3	Mercury

Secondly, we must take into account the influence of the sun's mass, which is put at 332,000 units.

The sun's diameter is 865,000 miles, and if we consider this as being positive and its mass negative, division by the latter 332,000 into the sun's diameter results in this:

$$865{,}000 \div 332{,}000 = 2.6$$

That is, 2.6 gravitational units. But there is another influence to be considered here, which is that atoms are being created in the sun. We think that deuterium is being created into helium in its central area. But we can only, in this instance, make a guess as to the figure of this. We can only make this fit in with other figures, so as to produce an actual amount. We have chosen, therefore, to estimate this amount to be half of the gravitational units. That is 1.3 units. So our estimate of the sun's rotation, assuming that 24 units equals one day, comes to this:

```
 24.9
  2.6
 +1.3
 ----
 28.8
 -3.1   Pi = 3.1 as a limiting factor
 ----
 25.7 days and hours
```

But our interpretation in these matters also includes this estimate — 25 days and 7 hours for the sun's rotation is approximately correct. But just like the earth, the sun has a definite cycle, which lasts for eleven years. After this period, that is, the sun's surface becomes cooler. We call them sunspots, and we reckon that, if combinations in its interior of deuterium were not taking place all the time, the sun would become totally dark at the end of its cycle. It would become a variable star, like so many other stars visible through telescopes.

It will be noticed that at the end of our interpretation figures (pi) — that is, one diameter and three figures for our circumference — is subtracted from them. This limiting factor should, we think, be applied to all massive bodies, for without it they would have no shape at all.

So we know, or suspect, that the sun was first formed from ice particles or some other kind of cosmic dust, and that subsequently interstellar gas (that is, hydrogen gas) became a large part of it. It fell onto the sun at the rate of a million tons a second. What colour was it then? It was a dark body, drifting in space within the galaxy. What happened then? Its internal pressure became such that the interstellar gas began to combine into deuterium, which atom is composed of one proton and one neutron. What happened was that, as a result of protons colliding, one of them had its field charges reversed, so that they assumed an inward direction.

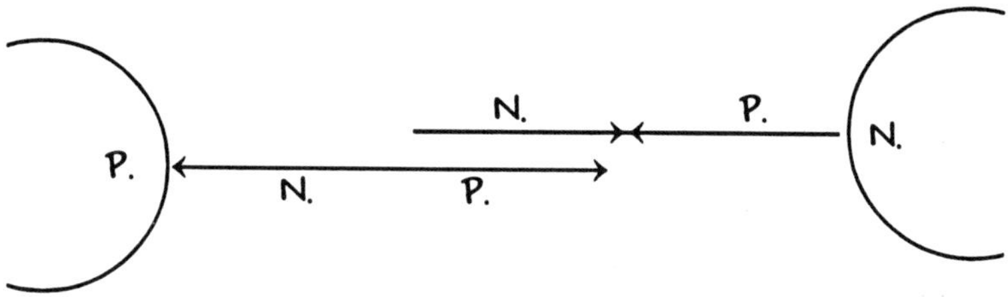

This union was very strong. As the sun continued to become more massive with interstellar gas falling on it, all the time more and more deuterium atoms were formed. These gave off emanations in the process. What had happened was this:

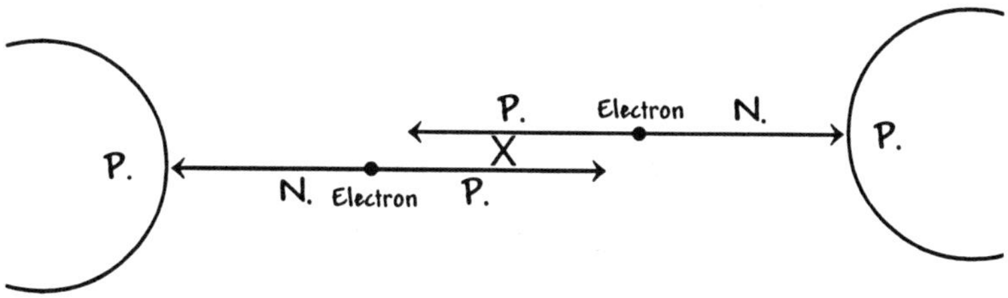

As the protons came together, their long-range positive fields were in collision, with the result that two electrons in orbit about them, coming under the influence of these positives, were thrown violently away out of their paths with some velocity. At the same time, one of the protons' long-range positives reversed its charge and became negative in its character, and to fit in with this its short-range charge altered its rotation and became positive in its nature. (It is noticed here that the atomic fields rotate as positive and negative magnets testify.)

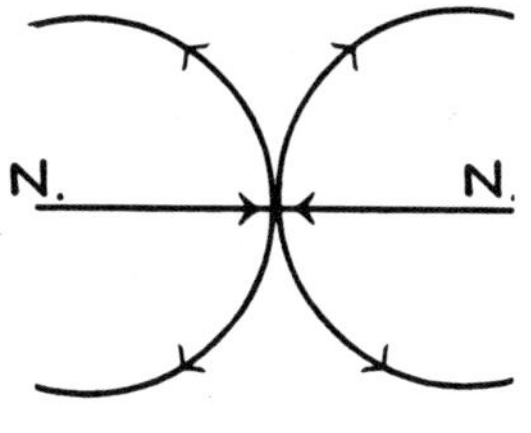

The disposition of Forces about a Negative Magnet

So electrons were thrown out of their orbits into the surrounding matter, this matter being, of course, other deuterium atoms. The situation was then that many electrons penetrated into the positive fields of adjacent protons which had lost theirs already, giving off, as they did so, electromagnetic radiations.

These were low-power emanations, the particles being so small as were the fields they passed through.

Now, the least in frequencies that an electron gives out are radio waves (domestic circuits put them out), and so we believe that, subsequent to the sun's creation of deuterium atoms, it began to generate radio waves — it became a radio star. This is hardly surprising when we are told by astronomers that there are as many radio stars in the sun's galaxy as there are giving out light radiations.

After ages had passed, with the sun becoming more and more massive, deuterium atoms began to combine in it, their electron emissions becoming then so powerful as to generate light and heat.

We think, therefore, from this stage that atoms in the sun were built up to the largest and heaviest ones, its heat causing it to assume an enormous disc. This process only came to an end with the creation of uranium and elements of a similar nature in it.

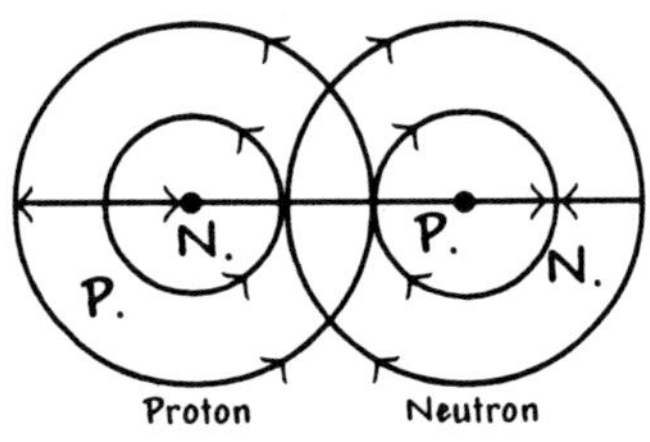

This is the skeleton structure of a deuterium atom. We can observe from this figure how its rotational forces are positioned. They are quite opposite and complementary. The dots in the middle represent the nuclei and the circles their fields. The outer circle of the proton (P) just touches the inner one of the neutron, while the outer circle of the neutron just touches the proton's negative. We think that it is the collision of the positives — just touching as can be seen — that prevents a direct approach between the nuclei of these atoms and maintains a distance separating them. They are joined together because they are opposites, the outer positive of the proton being held by the inner-directed long-range negative of the neutron field forces of a free deuterium atom. It has a single electron rotating around it, attracted there by the proton's positive nucleus.

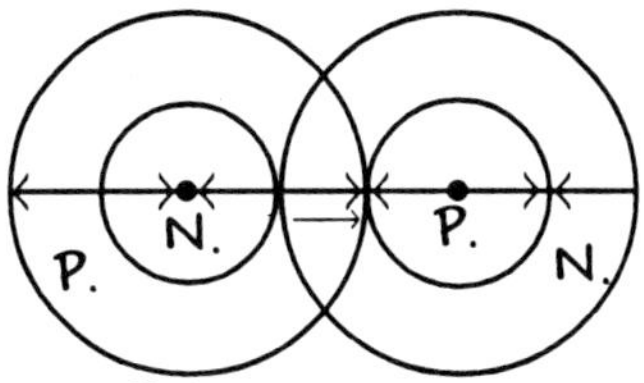

In this figure the direction of the field forces can be plainly seen. The positives take an outward direction, while the negatives assume an inner one. As a result of this it is obvious that a proton has a long-range positive field while a neutron has a long-range negative one.

It is a very well-known fact that electrons give off radiations. In attempting to find out why this is, we should first look at the electron structure. Although its twin forces are inwardly directed, they are composed of opposite charges, positive and negative, and this fact is certainly the case with atomic fields.

Now, we know that positive forces always produce motion in matter, and scientists are aware that as electrons assume orbits about atoms they have first to pass through their long-range positive fields. Thus there is a positive collision resulting. This comes about because the negative electrons first absorb positive force and then vibrate, giving off motion in definite cycles. It is the number of these cycles which determines the character of the emanations.

Thus the sun became a radio star, and it was from that point that larger atoms were created in it.

It used to be thought by the world of science that it is the fields of atoms which composed space. This was 'proved' wrong by experiments, but these, we imagine, were faulty. How else can electromagnetic rays travel if not through some medium such as is radiated by matter? Such experiments took no notice of that field radiated by the sun, which keeps the earth orbiting about it at the rate of 18 miles per second. We can't see it or feel it but, just like the force of gravity, it is certainly there. Even Einstein admitted that he did not know what light was. Apparently he did not know either what radio waves, X-rays or radar were. One scientist even said that while Einstein made clear several things, in others he had "confused the issue". Light, according to us, is just like a collection of ripples in a pond. Positive space is the medium. If this were not so, there would be no universe at all.

Why does light always travel in a straight line? The simple answer is because there is nothing to stop it. It can be diverted — refracted — but always resumes its straight course.

The original movement or rotation of the negative particles or electrons about their parents (the nuclei) begins like the planets do about the sun — that is, in a horizontal plane.

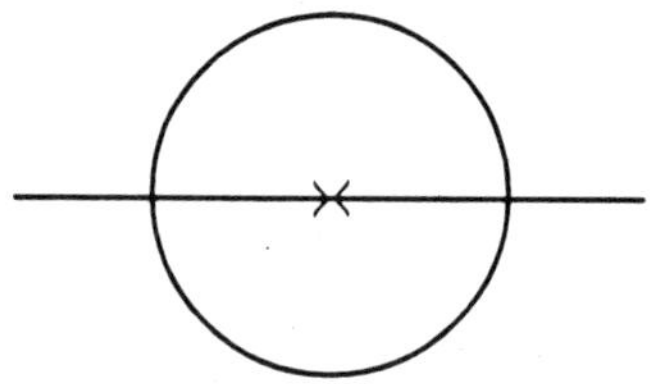

As these nuclei are augmented, however, this arrangement changes, but not as indicated in this figure, which represents tritium and is unstable, and which consists of a triangle of one proton and two neutrons.

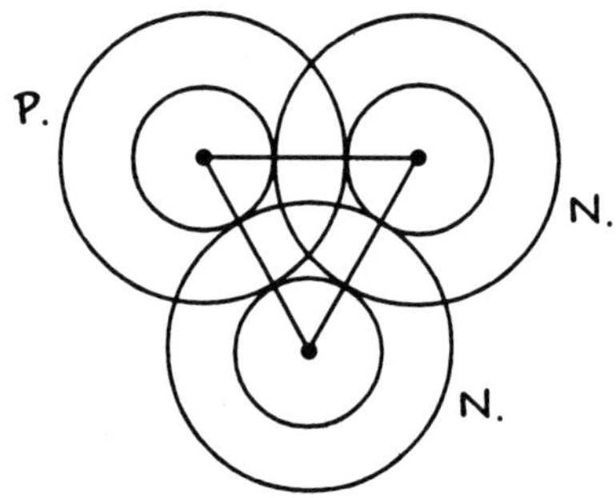

One of these neutrons tends to form a proton, the result being that two electrons are thrown off, making this atom radioactive. The next figure is different. This is the alpha particle, helium.

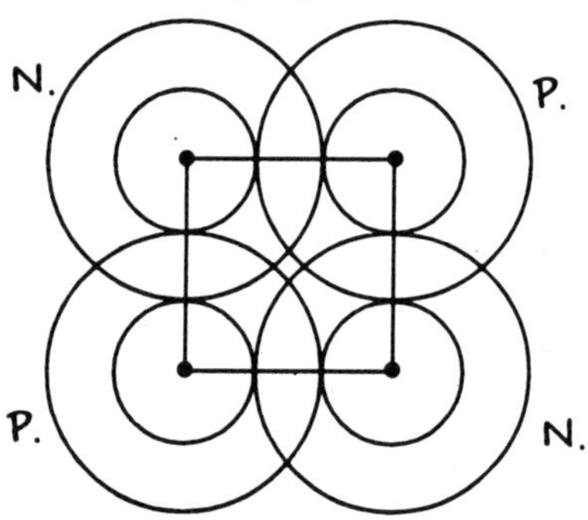

It consists of four particles, probably forming a square. It is also inert

— that is, it cannot form unions with other atoms. This figure shows the positive particles just touching each other, and it is the same with the negatives.

But what causes this inertness? We think that it is caused by a rotation of forces inside the alpha.

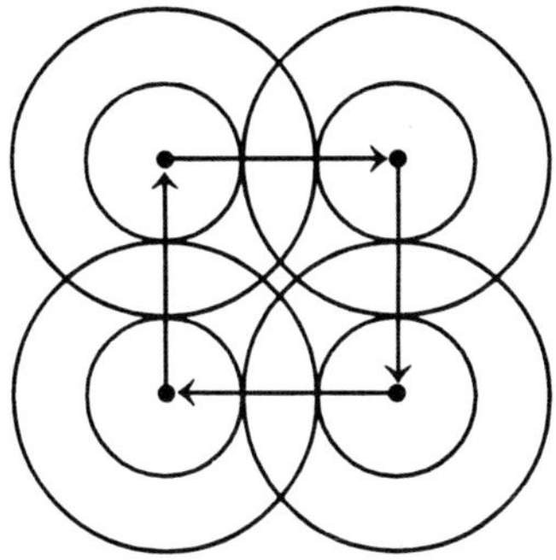

The protons pass on their outward-moving forces to the neutrons, and these then pass them on to the other positives, and so on. We can see this in iron magnets. When the opposite charges of these are put together, the magnetism is lost or is greatly reduced. A state of inertness governs them and if another piece of iron (called a keeper) is placed at the ends of the oppositely charged magnets, their conjoined internal forces will be maintained for a much longer period than if they existed singly. In these cases it is plain that a rotation is taking place between the two magnets, and the resemblance to helium or alpha particles is complete.

The nucleus of atoms is made up of alpha particles and, since they are inert, only their central or component nuclei remain positively charged, and it is these multiplied which generate atomic energy.

But what about the electrons? If they possess positive force they can move at the speed of light, and two of these attracted by the positively charged nuclei of the protons in the alphas rotate about them. It is always the law in the scientific world that like charges repel each other, and this we think happens in this case — the two electrons are repellant to each other. This happens because they are both in receipt of a strong positive force. They tend to collide when they are in orbit, and the only way they can avoid this is to rotate at and angle to each other. We can compare this process to the earth's rotation and its magnetic poles — they are hostile to each other.

This angular disposition of the electrons about the atomic nuclei is maintained throughout the extent of the atomic table. They are very close to one another because the multiplication of the protons' nuclei advances at a very slow rate, and it is this which gives their orbits to electrons.

Part 2

The sun and the earth are closely related. They are made of the same elements, mostly iron. Why iron? Well, we think that is because that element easily bears a positive charge, the outer half of which is repellant and resisted the iron atoms being combined with the deuterium atoms when elements were being created in the sun vast ages ago. So we think it is quite possible to equate the earth with the sun, the results being almost conclusive.

If it is taken, as is described later in this sketch, that both the sun and the earth have heat cycles, then these figures become applicable: earth's moon and its distance produce heat in the planet. Coupled with the sun's eleven-year heat cycle, this gives us by division:

sun's heat cycle — 11)238,000 — distance of the moon from earth

21,636 — degrees

Now, if we take this result and divide it by 25, that being the number of days it takes the sun to rotate on its axis, indicating its pressure and heat, we get this:

25)21,636

865

So this division gives us the number 865, and if we equate this with the moon's 238,000, we have 865,000, which is the sun's diameter in miles. We presume from this that if we multiply 865,000 by 25, we obtain the sun's heat in degrees at its centre at the peak of its cycle — a figure one thousand times greater than that obtained in the earth.

$$\begin{array}{r} 865{,}000 \\ \times 25 \\ \hline 21{,}625{,}000 \end{array}$$

From the foregoing we can infer that at one period long ago the sun was much hotter than it is now and that it has shrunk to a fraction of the size it once was. What its temperature was in the remote past before its family of planets was formed, we can only guess at, but it is obvious that it must have been extremely great.

Other things considered, it seems to us that its temperature must have been in the region of 865 million degrees, and adding its 21 million present temperature we have a figure of 886 million degrees, which is the distance in miles of the major planet, Saturn, from the sun.

$$\begin{array}{r} 865 \\ +21 \\ \hline 886 \end{array}$$

Clearly what happened to produce these figures was that the sun expanding a thousandfold had formed a great disc from its equator, extending as far out as Saturn. This it did as elements, up to the heaviest, were being created in it.

Now we have to guess again as to what took place after the disc was formed. Then, by the influence of the inverse-square law, more gas (and it is known that Saturn is mostly composed of that substance) was thrown beyond that distance — 886 million miles — to distances far beyond that figure, and, as this cooled and formed into separate bodies, these all began to exert a brake on the sun's rotation, which is the situation being maintained today.

This braking influence must be considerable and exert a great influence, both on the sun's rotation (25 days) and its heat cycle, which takes only eleven years to complete, compared to the earth's 238,000 years. (This is the distance in miles of its satellite, the moon.) This is pure conjecture, of course, but it does fit in very well with scientists' estimation that there have been four ice ages, interspersed with warmer periods, during the last million years.

It may be asked of the earth's cycle and also the sun's: How do they come about? Well, in the earth's case it has been remarked that it and the moon are more like a binary combination than ordinary planets and their moons. But the first thing we must explain about their relationship is that the moon exerts a brake on the earth's rotation. We all know that if a brake is applied to a spinning wheel, the wheel slows down and the brake becomes hot; this, in a roundabout way, is exactly what happens in the earth in response to the moon's brake. The sequence goes like this: first, on the moon's brake being applied the earth's rotation slows down. This causes increased pressure at its centre. Atoms there (almost certainly iron atoms) forced together, tend to lose their outer electrons, thus bearing positive fields. This causes the earth to become magnetic — great poles are formed which, starting on the earth's equator (that is where the most motion is lost), tend because that rotation is positive (all motion is positive) to be deflected away from it. One magnetic pole is deflected north and the other south so that finally these magnetic poles reach the geographic ones — that is, the furthest they can get away from the equator. But, at the same time as this is happening, the earth's pressure at its centre being increased, it begins to rotate faster. This happens slowly, of course, until such a point is reached to allow electrons to begin to return to their orbits, as the relaxation occurs at the centre of the earth. Now, it is very well known that as electrons resume orbits they give off radiations, these being normally light and heat. The light does not affect the earth but the heat does — it starts to become warmer, and at the same time the returning electrons begin to cut off the earth's magnetic poles. They start to fade and, their positive force becoming less, they slowly begin to move back towards the earth's equator. This they do, becoming weaker and weaker until finally they disappear when the equator is reached.

The earth's production of heat at this stage is at its greatest, all its central electrons having slowly resumed their orbits. But what happens then? The moon's brake is still being applied, so the whole process of the production of heat in the earth, caused by its slowing, is repeated. Its magnetic poles appear again, moving away from the equator until finally, again, they reach the geographic ones. This whole process we can call the earth's cycle — it becomes warmer and colder alternately. The situation of its magnetic poles acts as a pointer to the progress of its cycle.

What do we see at the present day? The northern magnetic pole is in the north of Canada, moving south towards the equator and fading, so we can expect that the earth is becoming warmer as the electrons at its centre return to their orbits. The earth's rotation is slowing down fractionally, but most of this motion, we think, is inertia. We can gauge almost exactly the stage of the earth's heat cycle at the present age by the fact that the northern magnetic pole is about 2,000 miles from the North Pole, and as the distance from this pole to the equator is about 6,000 miles we can see that the earth's heat cycle is about a third completed. It is just emerging from a cold period (an ice age) and is becoming warmer. Scientists have estimated that the ice at the Antarctic continent has melted by about a third since the last ice age, so we think our theory is pretty conclusive.

We mentioned at the start of this discourse that the period of the earth's cycle is about 238,000 years, as compared to the sun's eleven cool areas, appearing on that enormous body every eleven years. This cycle is caused, of course, by its satellites' brakes. We can obtain the earth's figure by using our imagination and supposing that if the moon was a mile from the earth, its cycle would be completed in a year. But as that satellite is 238,000 miles from the earth, we can guess that that figure could represent the length in years of its cycle. It might be asked at this juncture why, the magnetic poles having vanished on the equator and the electrons having all returned in the earth, is that where the earth is at its warmest? The simple answer is that from that point the earth begins to cool and this process continues until the magnetic poles reach the geographic ones.

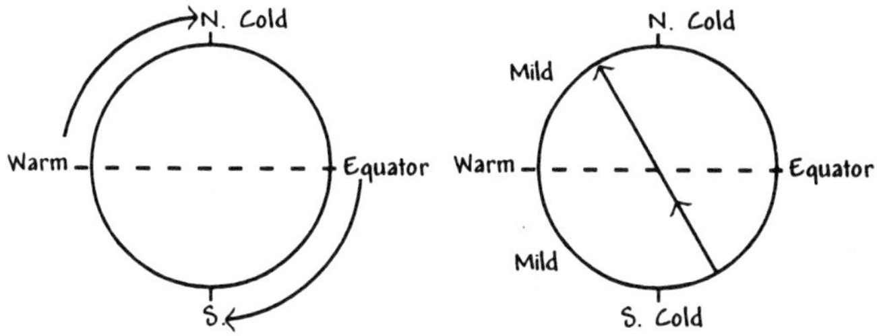

Figure Showing the Present Positions of the Earth's Magnetic Poles

The sun swops its magnetic poles round every eleven years in exactly the same fashion as the earth does in approximately a quarter of a million years.

The present conditions, however, were not always like this. Millions of years ago scientists tell us the moon was much closer to the earth than it is today. The earth's cycles were much quicker, merging in one another, so that instead of ice ages the earth had just cooler periods from time to time, and in those conditions life flourished. Cold-blooded dinosaurs roamed the earth as did many other forms of life, as fossils testify. Gradually, though, as the moon receded and the earth became cooler, cold-blooded animals were replaced by warm-blooded ones which could stand the cold. These included man, who probably hunted several species to extinction, including less brainy men of his own kind. So we come to modern man and his increasingly warm earth but, however clever he may become, he won't be able to do a thing about his climatic conditions.

As regards all the foregoing it is popularly supposed nowadays, and actually believed by many scientists, that the present global warming is caused by the earth's ozone layer causing a greenhouse effect on it. But what are we to think as at the same time geologists have estimated that the ice on the Antarctic continent has melted by a third since the last ice age. No reason has been given for this and some people think, as did Professor Fred Hoyle, that the sun is responsible. We don't accept this. Isn't it a remarkable coincidence that the earth's magnetic poles have moved a third of the distance (2,000 miles) from the geographic ones in the same period of time? We think our theory is the correct one.

Taking another cursory look at some of the figures of the foregoing, we can start with the earth. We made it out that its entire cycle took place in 238,000 years, and that if we divided that figure by 11 we obtained 21,638 as a result. We then assumed that if we changed that figure to degrees, that was the temperature being maintained in the earth, and, iron melting at a temperature of 1,500 degrees centigrade, we supposed that was the reason for there being many volcanoes and earthquakes on the earth's surface. What we then did not appreciate was that the number 11 had a special significance in that if this figure was changed to degrees it would indicate the change in temperature it would have as a result of the earth's 238,000 years' cycle. If we do this, we can plainly see that that cycle change of temperature would be eleven degrees at its highest, that total occurring when the magnetic poles were at the equator, these having faded to hardly anything.

Again, if we took the earth's 21,638° and divided them by the number of days the sun took to rotate on its axis once, we had this:

$$\begin{array}{r} 25\overline{)21,638} \\ 865 \end{array}$$

That is, 865, which being increased a thousandfold to equate it with the earth's thousandths, would be 865,000, which as we know is the sun's diameter. We now have no reason to change this, as it seems to us that we were dealing with actual facts and not theories.

Also, if we multiplied 865,000 by 25 we could reach the figure of 21,625,000, which we suspected was the sun's internal temperature, its cycle being completed in eleven years.

$$\begin{array}{r} 865,000 \\ \times 25 \\ \hline 21,625,000 \end{array}$$

Also, if we take now the sun's surface temperature to be 10,000°F, and its cool period at the end of its cycle to be 2,000 degrees cooler, that gives us 8,000 degrees for its sunspots. This corresponds to the earth's ice ages. We can notice also that 8,000 is the earth's diameter in miles.

So what about the coming warmer period of the earth? This will take place over many thousands of years until 11 degrees is reached. It is obvious that in the future the earth will become a great deal warmer, but nothing like as warm as might be anticipated. It must be remembered that the earth's surface is covered — 75 per cent of it — by water, sometimes at a depth of several miles. Now, water is a great absorber of heat, so that instead of many areas of the earth experiencing greatly increased temperatures, these will not happen. What will happen, we think, is that the earth's air will become more humid. This will lead to an increase in the number and severity of electrical storms on it, which is actually what is taking place now in several areas of the world. Monsoon rains will become heavier too.

But there is another side to this — moisture in the atmosphere, affected by increased heat, will evaporate so that large areas will experience drought conditions, which, in fact, is already the case in some continents.

Another circumstance in regard to this concerns the earth's present heat. If the maximum of this is going to be eleven degrees, then the present one, since the earth's cycle is a third accomplished, is easy to find out by dividing 11 by 3:

$$\begin{array}{r} 3\underline{)11.00} \\ 3.66 \end{array}$$

The result (3.66) seems to indicate that the earth's surface temperature has risen by this amount since the last ice age. This, of course, does not seem very much until we realise that a temperature of one degree above freezing point can melt the largest of glaciers and icebergs very quickly. All the signs of this are going on now, at both the North and South Poles. It is stated that, in a comparatively short time, all the ice at the North Pole will have disappeared so that ships will be able to sail across it without any difficulty.

Part 3

The crust of the earth, geologists think, is about 60 miles thick and divided into plates which ride on semi-molten material, almost certainly mostly iron. Where these plates join, volcanoes and earthquakes occur, which effect is not surprising when we consider that the earth's internal heat can be put at 21,000°C and so can be compared to a boiling kettle from which steam is escaping, and bearing in mind that the melting point of iron is 1,500°C we have a picture of violent eruptions of volcanoes occurring from time to time. One of the most violent of these, Krakatoa, took place in the nineteenth century, when the top of the volcano was blown completely off and fell 50 miles away forming a new island.

Another effect of the earth's great internal heat concerns what is called continental drift. Its surface plates are in constant motion, so that with the passage of millions of years continents and islands have changed their positions considerably from what they were in the remote past. It is our theory that most of this movement is caused by the basic influence of the earth's satellite — the moon. We imagine that the braking action on the earth's rotation causes heat to be generated in that body, and this welling-up has the power of moving continents as well as warming its surface.

We saw previously how the moon's drift from the earth was measured in inches, and the same, we think, is probably true applied to continental drift. We can call this rate of movement, therefore, approximately one inch per annum — this figure being an average one.

200 million years ago, the geologists tell us, all the earth's continents were joined together into one great mass, which they call Pangaea. After a considerable period, however, due, we think, to the moon's influence, this mass split up and two continents were formed — Asia and America. If we then take 200 million inches and convert them into miles, we find that they add up to 3,156 miles (approximately the width of the North and South Atlantic) and in addition to this we find that the same rock substances are found on both sides of the Atlantic. So we might be aware now of some changes that have taken place on the earth's surface during the last 200 million years. But, first a few more figures: Multiplying the diameters of the moon and the earth — 4,000 × 4,000 = 16 million for the earth, and 1,000 × 1,000 = 1 million for the moon — that is 16 to 1. But this is not the correct estimate because the earth's negative mass is put at 80. If we divide 16 by 80 we get 0.2 or one fifth, so the ratio between the earth and the moon is in reality 5 to 1.

80)16.0
.2

The moon's 1 is 1 million, so we can say that the moon exerts a fifth influence on the earth, more or less (more in the past, of course, because the satellite was more close to the earth then).

Now, there have been, during the last 200 million years, five most important changes of the land on the earth's surface, and, as regards nearly all of them, the distance or its effect can be estimated to be 3,156 miles. First we have the North and South Atlantic's continental drifts, with volcanic Iceland in the north and, similarly, volcanic Tristan da Cunha in the south. Then we have Alaska, part of North America, which it is thought by geologists to have been in the past several thousand miles further south than it is now, as is evidenced by there being both coal and oil there — the products of vegetation. This observation can also be applied to Antarctica, which continent also contains coal, which suggests that millions of years ago it was much further north than its present position. It also has volcanoes on its surface which, we think, in time encourage vegetation to grow there (volcanoes always seem to encourage fertility).

We come now to the Indian subcontinent, but it wasn't always like that. At the time of Pangaea, India was an island continent. Then it gradually began to move north. Starting from the seabed, it slowly pushed the nearest land mass of Asia into first high ground and then into mountains. These have become the highest in the world — the Himalayas. The highest is Everest, which is 30,000 feet. How do we know that these started from the seabed? Because the fossils of fishes and even whales have been found in them.

The last country which we will consider is similar to India but much smaller. It has active volcanoes and, like India, was ages ago an island which, slowly moving north and colliding with the mainland, begat a mountain range. This country is Italy. Such was the force propelling it north at one time that, even though it was 15 times smaller than India, the Alps it produced were only half the altitude of the Himalayas, Mont Blanc being 15,000 feet high.

What changes will take place in the future we can only guess at. Certain it is that as the moon continues to recede the earth will become colder and mountain-building will stop or slow down considerably, as it has done in the modern age.

The effect of the moon's brake, 5 to 1, will continue to have some influence, of course, but nothing like that it had on the earth's surface in past ages.

CHAPTER 2

Part 1

Professor Fred Hoyle, in his book, *The New Cosmology*, which was widely read and broadcast and came out after the last war, remarked that most of the world's scientific problems had been solved, but that the fact of gravity about masses was still a mystery to him. We have never read in our sixty years of reading any viable explanation of it. We have one, and it goes like this: When for some reason electrons about atoms leave their orbits, the atoms are said to be positively charged — that is, they then have a short-range negative attractive charge and a long-range positive repellant one. These, we believe, came into being, together with the negative electron, when the atoms were being created. Pressure was then exerted on their nuclei, forcing fields on their equators to be formed. So the actual set-up was like this:

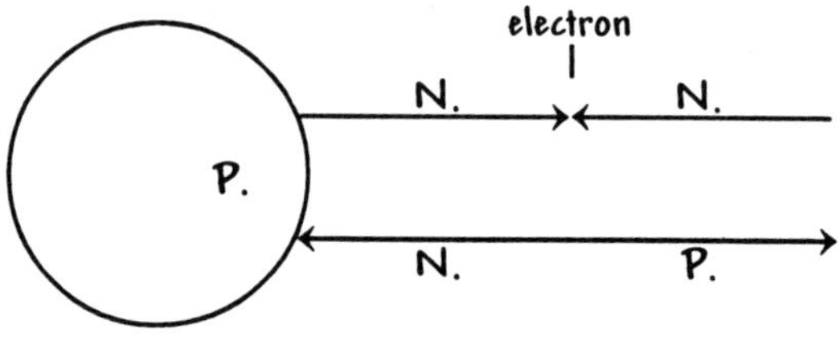

As we can see, the negative force produced a particle and the positive one two charges — one an inwardly directed force, and the other an outwardly directed repellant one. The negative particle, when in orbit, being exactly the same strength as the two positive ones, cuts these off and replaces them with its own.

When, however, it leaves its orbit the two basic ones are revealed. Now we come to the conclusion that the basic surrounding fields of all matter are the sum total of their positives, and, this being so, they have a short-range attractive force and a long-range repellant. We think, therefore, it is the short-range attractive force — half as rare as the electron, the smallest particle, which exists in billions in matter — which produces what is called gravity. This force tends to induce all matter near it to move towards the centre of the mass producing it.

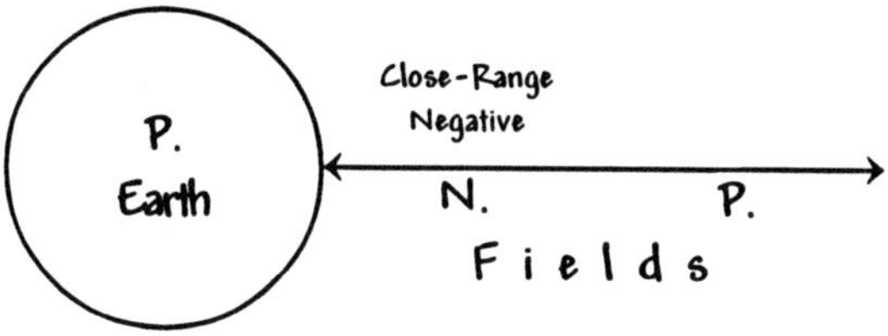

On the earth, which we believe to be a great iron ball (and all metals tend to be positively charged), this negative inward tendency is so strong as to require a velocity of 24,000 miles per hour to escape it. This is achieved, of course, by modern rockets, which can exist as satellites. Some last for long periods, but others, influenced by the earth's short-range negative attractive field, gradually approach the earth until they burn up in its atmosphere.

So if a man jumps up at the North Pole, the earth's negative immediately pulls him down again. Just the same happens at the South Pole, but what, according to us, does the other half of the earth's field, the long-range repellant, do? If it has one, does it repel matter? We think the earth has one and it works in this way: The moon, the earth's satellite, orbits it in its repellant field some 238 thousand miles away from it. It is in receipt of the earth's repellant positive, and this, colliding with its own internal positive, causes two things to happen.

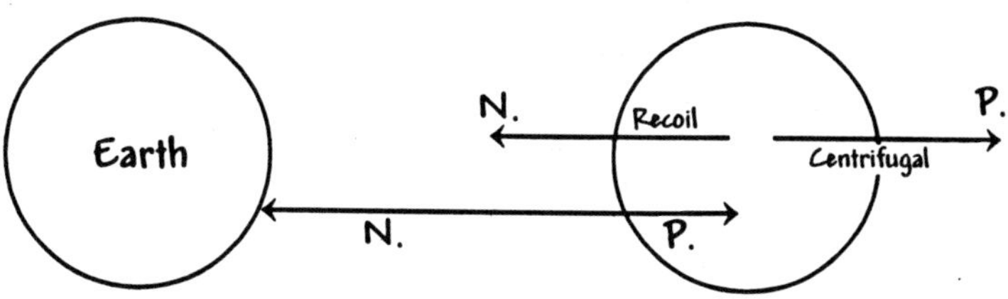

First, there is a recoil which impels it towards the earth and, second, there is a repellant positive which influences the moon in the opposite direction away from the earth. This is what scientists call a centrifugal force. The moon, influenced by these two opposite forces, obeys both of them. First, it moves — falls, recoils — towards the earth and, second, its positive acting in the opposite direction causes it to maintain its distance from that body. In short, it rotates about the earth at something like 2,000 miles per hour. It used, we think, to travel much faster than that when millions of years ago (about 200 million, we believe) it was much closer to the earth.

Part 2

As soon as a considerable-sized rotating mass obtains a fair-sized satellite, the latter exerts a braking influence on the former. This depends on distance. What happens then is that as the larger mass's rotational centrifugal force is weakened, so its central pressure increases. This tends to quicken up its rotation. As this happens, centrifugal force comes into play again, pressure is relieved and the mass slows down. This process, repeated all the time, causes the mass to have an average rate of rotation. This takes place normally over long periods of time.

It can be compared to the action of a governor on a steam engine. This consists of a valve through which steam can pass. Attached to this valve are two weights, which, when steam pressure forces them up, can close the valve and thus shut off the flow of steam. A steam engine, equipped with a governor, can therefore run at a constant speed because the flow of steam passing through it is always being regulated. The two weights allow steam into it — as the pressure drops, the weights open the valve.

We all know that the earth's forces are such that, at a distance, masses rotate around it. To explain this fact the world of science seems to imagine that a force, called centrifugal force, prevents a direct approach of rotating masses. This force appears (or seems to appear) almost like a rabbit out of a hat. No real reason — no *real* reason — is given for it. As regards the earth's close-range negative existence being proved, suffice it to say that it requires enormous force (escape velocity) to overcome it and put satellites into orbit. These rotate at great speed around the earth a few hundred miles up, but they don't last long. Those furthest out, because of the weakened negative, survive the longest.

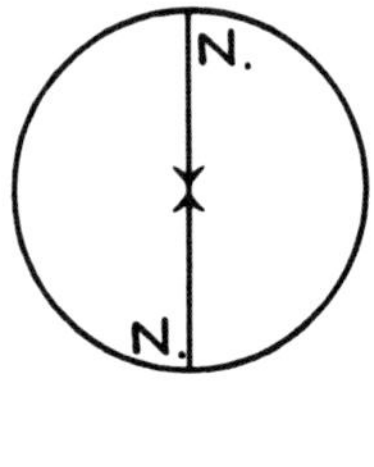

Earth's Central Pressure

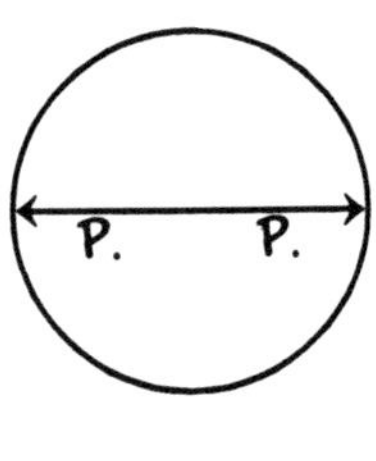

Earth's Diameter and Rotation Relieving Pressure

The strength of the earth's negative can be gauged fairly well if we look at a few figures. It is 8,000 miles from pole to pole, so, dividing this and multiplying 4,000 by 4,000 gives us 16 million. We can take this as being the earth's central pressure, and it is this forcing positively charged atoms together which produces the positive force of motion. Looking at the second figure we can see how its diameter and movement can relieve that pressure. Balancing these two separate forces, we might think we have the earth's rotation, but scientists inform us that the earth's negative mass is computed to be approximately 80 units, so we must now make these figures:

4,000
×4,000
Mass 80)16,000,000
Diameter 8)200,000
25,000

As a result of these, we can arrive at the figure 25,000, and it is this we can assume to be the strength of the earth's gravitational positive force at its surface.

We all know, however, that the earth's circumference at its equator is 24,000 miles, and so subtracting this figure from the positive 25,000 leaves us with a thousand positives and if we assume this to represent a thousand miles per hour, we can now see how it is that the earth's rotation on its axis at that velocity gives it a twenty-four-hour day.

Every mass on the earth, such as chairs, tables, etc., has its own tiny positive and negative fields, and it is these, coupled with the earth's great negative, which produce gravity which keeps them very firmly on the ground.

The inverse-square law plays a great part in the earth's gravity. Eight thousand miles up from its surface (8,000 miles being its diameter) it is reduced to a quarter of its strength, the negative then being only 6,250 units.

4)25,000
6,250

This is the reason why astronauts in space find that their weight has gone down considerably compared to what it was on the earth. Motion in conjunction with the 1,000 miles per hour maintains the earth's 25,000 positive strength.

8,000 miles away from the earth (that is its diameter) the earth's positive, producing the motion of its satellites, can be, as stated, 6,250 miles per hour. The motive units generate that velocity.

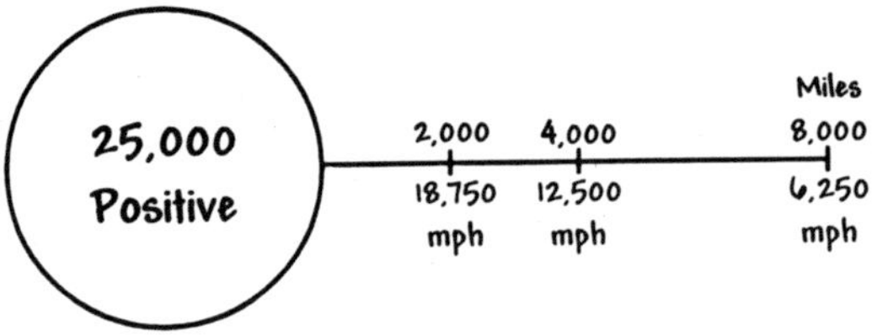

Four thousand miles away a satellite's speed would be 12,500 miles per hour, double the speed at 8,000 miles distance. At 2,000 miles distance a velocity of 18,750 miles per hour would be attained.

$$2\overline{)25{,}000}$$
$$2\overline{)12{,}500}$$
$$\text{Furthest} - \underline{6{,}250}$$
$$\text{Nearest} - 18{,}750$$

It may be mentioned here that the figures given are very close to those experienced by astronauts in their rockets when they are orbiting the earth. These, under the influence of the earth's negative gravitation, often travel at about 17,000 or 18,000 miles per hour at distances of a few thousand miles above it.

If we continue with the earth's mass and its rotation we should take another look using a decimal equation:

$$80\overline{)16.0}$$
$$0.2$$

This equals two tenths, which is, of course, one fifth. It is gained by dividing the earth's negative mass into the result of its two fours multiplied, these being its 8 diameter (= 4 × 4 = 16) and the earth's mass, put at 80. We should now, therefore, divide the resultant fifth into the earth's twenty-four-hour rotation, obtaining 4.8 gravitational units:

$$5\overline{)24.0}$$
$$4.8$$

We should now, however, take into account the earth's satellite, the moon, and since we are dealing with simple figures and the moon's diameter is 2,000 miles — a single unit, a fifth of the earth's figure — five can be achieved. The earth–moon relationship can be, therefore, put at 5 to 1. But we have already seen that the earth's internal figure is one fifth, so how can we square these two figures?

We can only assume that one fifth is internal and the other external. The moon influences the earth's external movement, while the other fifth, its internal one. They are complementary. The internal one governs its rotation on its axis, while the other governs its outer rotation and tides, and the movement of the plates on its surface together with earthquakes and volcanoes.

We have seen that movement deep in the earth caused a rotation on its surface of 24 units, and in the previous equation we saw how, by dividing this by five, 4.8 resulted.

If we take 24 units as a base, we think it is possible that we can also find out, by addition and subtraction, the number in hours of some changes which have occurred on the surface of the earth in past ages.

But first we must consider how long ago the moon became attached to the earth, and what were the conditions then. We think the earth acquired the moon about 200 million years ago, and since it has always been in its repellant field it has very gradually moved away from the earth, which was not its parent.

So, using our figures, we can discover the earth's rate of rotation, which it had all those years ago. This is simple.

$$\begin{array}{r} 24.0 \\ -4.8 \\ \hline 19.2 \end{array}$$

From this figure, it is plain that before it acquired the moon, the earth rotated in 19.2 hours. Similarly, still using the earth's twenty-four-hour rate of rotation, we can find out what the earth's rotation was immediately after the moon became attached to it. We must remember first, in regard to this, that at that period the moon was closest to the earth, it having receded since that time:

$$\begin{array}{r} 24.0 \\ +4.8 \\ \hline 28.8 \end{array}$$

We can observe from this figure that less than 200 million years ago the earth's rotation was approximately 28.8 hours. Since then it is obvious that as the moon went away so the earth's rotation increased until it reached the present rate of twenty-four hours.

Heat in the earth has a great deal to do with its rotation, but nothing like that in stars. In them their heat reduces their pressure-producing centres, but in a cool contracting mass heat can increase its internal pressure, causing more rapid rotation. We can see this plainly in the sun and its planets. The major planets, being cool, rotate at a great rate, while the sun, being extremely hot, rotates very slowly.

CHAPTER 3

Part 1

Like nearly all animals and plants, man has been afflicted by parasites and diseases. We consider the most important disease he has suffered from is that which mostly affects the lungs — namely, tuberculosis. This disease, caused by an acidic bacillus, is most often contracted early in life from fine droplets from another person, usually a relation. The conditions of its contraction are most often the same in that person. It is the case, therefore, that this disease tends to run in families.

Its advance is so slow as to be unnoticeable, and it is characterised by coughing and night sweats.

It is generally in the teens or late teens that the disease becomes evident, almost always affecting the lungs. Many people contract this and get over it without knowing they have had it. Their physical condition is not conducive to its development.

The lung tissues, when affected, react to the bacilli by producing protective cells that engulf the disease organism forming tubercles. This disease is of a rodent nature — the acid bacillus eats its way into the lung tissues, releasing their binding electrons. Once a certain stage is passed and both lungs are affected, the disease ends unfavourably for the patient. However, in many cases the progress of the disease is so slow that the patient may survive into old age, although incapacitated in some degree. In some instances, though halted, the effects on the physical and mental processes of an individual may remain throughout life. In some case, as well, this disease may skip a generation and then reappear in the offspring thereafter.

Now, tuberculosis, by its strongly positive fields, tends to draw out the negative electrons of, in this case, the lungs, and by doing so it weakens them.

What happens to these freed electrons? We think that they circulate in a patient's bloodstream and being, so far as it is known, the smallest particle they can penetrate into every part of his or her circulation, including the genital organs.

As they are so small they can also affect the genes and chromosomes there. This means they can have an influence on any future offspring the patient may produce.

Electrons, by colliding with other electrons, tend to throw them out of orbit, rendering the atoms about which they were orbiting positively charged. How can this affect the genes and chromosomes? We believe that the offspring they originate vary both mentally and physically. Some will be strongly charged and vigorous; others will be weak in some way, being overcharged.

There are, of course, many variations of these two extremes and we have only to look at the offspring of tuberculosis sufferers to find this out. Mental instability, epilepsy, club feet, humpbacks — these are some of the conditions which occur in the offspring of tuberculosis patients. But it is the variations of mental instability which are most notable.

What is insanity? The answer to this must be that the person is abnormal mentally in some way and cannot act reasonably. In extreme cases he cannot do anything right. Then there is a type of this when a person has periods of insanity, and can have periods when he is alternately abnormally active and abnormally inactive. Such a condition is known as being manic. Then there are persons who are considered to be slightly insane. Eccentric is the word used to describe them. After this there are normal individuals who can be stupid or intelligent. Finally, there are clever persons or very clever persons and there is no upper limit to them.

All the foregoing can be the result of the action of tuberculosis, and it is the very clever individuals who make the most impact on communities, and have done for many thousands of years — ever since man appeared on the earth.

It is obvious, going by Darwin's theory — and, by the way, his daughter fell a victim of 'the English disease' — that in all history the cleverest would survive. Neanderthal man, who had a brain two thirds the size of modern man, has been extinct for thousands of years due — there can be no doubt — to wars between the two types of Homo sapiens. It appears to us — and there can be no doubt about this either — that only man's brain has been altered and improved since the time when he was a semi-naked savage. His body characteristics are the same as he possessed then. We can list a few of these surviving from the time when he was a fighting and hunting man, but first it may be remarked that from his earliest period he domesticated wolves, which by breeding became dogs. These dogs assisted in his evolution. They helped him in his wars against other tribes. The men left some dogs behind to protect their wives and children. To bind this association man retained part of his hairy coat, most of which had been lost in order to blend his colour with his surroundings. It doesn't do to be dark-skinned in a snow-coloured landscape. Similarly, it was a great advantage to be black in Darkest African jungles.

Dogs, as is very well known, live in a world of smells. They explore everything with their noses, and so it happened that primitive men retained scent-retaining hair on their bodies. This bound the dogs to their masters during the times when they went out fighting and hunting. Hair occurs in the armpits and on the chests of males, and around the genital organs of both sexes — places where they sweated most and had the strongest odour.

Men developed beards and moustaches at puberty, the former to protect the throat, and the latter to divert blood from the nose to prevent it from entering the mouth and interfering with his war cries, for which — by his voice breaking — he had developed a great power.

Females retain their high-pitched voices at puberty, these being heard at a long distance in times of danger. In the same way, of course, as those of children.

We think that, as well as producing scent, pubic hair on both sexes softens intercourse.

Sweaty feet have an advantage too. As well as binding dogs to their masters, the dog can trace him if he is lost or injured in some location.

Another noticeable thing about humans is that males tend to go bald or grey as they get older. The baldness begins at the back of the head, always at a spot where, according to phrenologists, a vital part of the brain's working comes close to the surface. After this the hair begins to recede from the front of the head — that area where the frontal lobes are situated. This process goes on in many males until all the hair on the top of the head is lost and the man is completely bald.

Women are quite different. They retain their hair throughout their lives, possibly to protect the skull from damage of some sort.

In men, however, baldness, or lack of hair, does not occur towards the back of the head. There it continues to grow normally, and we think that the reason for this is that it would serve as a protection for the back of the head by warding off sunstroke, primitive men being unable to prevent its growth.

What can we make of all this? First and foremost it would appear that in those savage days men needed their brains more than women in order to survive. A man had to have a shelter of some kind. The first ones would probably have been caves, because at that time man hadn't learn to construct his own shelters. Then he had to feed his family and protect it. He would need to fight other tribes and wild beasts. This would always need brains and we imagine loss of hair on his most vital parts (the brains) would keep them cool so that they would work better.

It is the same all over the world's various races, suggesting to us that they are all descended from a few similar tribes.

But what about the grey-headed men and many women (for many women also become grey with age)? Grey is a light colour and we think it serves the same purpose as baldness does. Light and heat are reflected from it, thus cooling, in some degree, the brain.

We must be reminded here of the blond and golden-haired races of the world. They possess a natural protection against heat.

It must be a simple fact, recognised by all races, that the eyebrows are there mainly to prevent moisture, blood or sweat, getting into the eyes and therefore interfering with vision.

There is, however, another significant feature in regard to human beings, and this concerns their mouths, or more particularly their lips. Lips vary in man from thick to thin. Why should this be so? Looking at man's neighbours amongst the apes — chimpanzees — we find that they have no lips at all. Has this a significance in some way? We imagine that it has. Chimps live in trees and they eat fruit, and we can compare their habits to the habits of people in different parts of the world, who live on fruit, vegetables and meat.

Eastern man lives mostly on rice and pigs. He has thin lips. Looking, however, at men from the Pacific area we find that they have thicker lips, live on coconuts and fish and, according to Jules Verne, Daniel Defoe and other men, they have (or did have before Western men went there) cannibalistic tendencies.

We know, however, as a fact, that in that area men have many animal enemies, which possess acute senses of smell. They are, for the most part, carnivores and can smell meat a good way off. If, therefore, a man ventured too close to one of these creatures with his breath and his lips putting forth a scent, it would very probably be fatal for him.

This is very much the case in Africa. There dangerous carnivores abound. The consequence, as we see in black Africans, is that they have very thick lips and also large teeth, the better, apparently, for eating raw meat of any kind. They possess, as well, very broad noses to enable them to breathe in hot rarefied air.

But why thick lips? Simply because, by licking them, traces of meat would be removed.

In considering these things, though, we must apply our theories to India. There big cats can be found, even more than in Africa. There are no tigers in Africa, but plenty in India — or there were until recently. But the lips of modern Indians are not noticeably thick. Why should this be so? We think that the reason is that Indians are not great meat-eaters. They like curry and rice. Indeed, there are few Indian dishes which don't contain these ingredients. Allied to this, there are in India many fruits which don't normally grow in the West and which the Indians eat. Thus a fair picture emerges.

There are many great plains in India, which are watered by the Ganges, the Indus and the Brahmaputra. The chief industry is agriculture. The main crops are rice, wheat and cotton, so we see there might be a reason for the Indians' appearance.

But switching our attention now to North America we notice that the Red Indians have ordinary thin lips. Before the white man went there the natives lived and worked with the buffalo and apparently ate its meat. There are no big cats on the North American plains. True, there are bears and wolves, but these live chiefly on sheep and smaller animals. A pack of wolves can outrun deer, but contrary to popular opinion these animals do not attack human beings.

About South America we have little information, but it appears that the natives, dark-skinned, lived mainly on fish and fruit, and closely resembled the North Americans.

Man, like almost all animals, spends some of his life asleep. Why should this be so? There was a commission set up some time ago to try and find out why. Their conclusion was that it was 'man's complete inability to stay awake'. Such a conclusion, we think, was completely ridiculous.

There are, we are told by medical experts, 10,000 million cells making up the human brain, and that each of these has a nucleus, surrounding matter and a cell wall, but our way of looking at these is different to that of the medical men.

We imagine that each of these cells is like a tiny magnet, having positive and negative sides to them as in this figure:

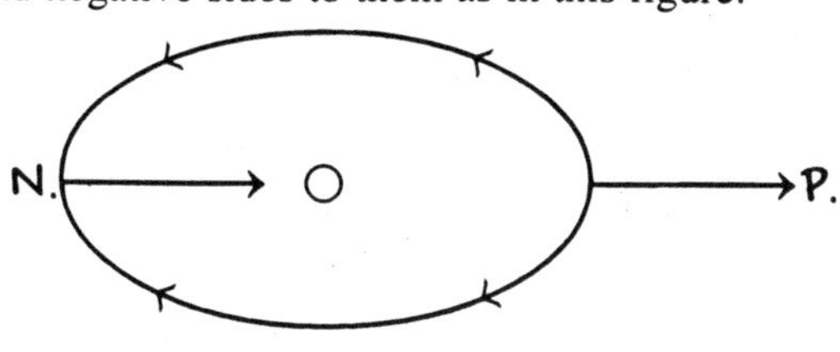

We imagine also that these during the day tend to lose their magnetism. This is due to their supplying power to the mental activity of the brain. They have to be recharged again, therefore, and this happens, we maintain, during sleep. If this did not happen, it is obvious to us that man, and most other animals, would certainly die.

What happens in sleep? We think that there is a sleep centre in the brains of animals which puts off electric currents all around them. These throw off other electrons which tend to collect there during the day, thus cutting off their positive charges. This process goes on for several hours and, after it is completed and the freshly charged brain cells are ready to begin working again, the animal wakes up.

We have already mentioned that there is a connection between a rodent disease and the central nervous system (the brain), and it is quite possible, due to the presence of electrons in the bloodstream, for many magnetic cells to become overcharged. When this happens the brain becomes unstable, and the affected cells give out strong electric currents. In the worst cases, this often causes loss of consciousness in the man or animal and spasmodic convulsive movements of the limbs. Such a condition is called epilepsy, and it can, in the most severe cases, cause death.

We can see now how a rodent disease can cause epilepsy.

But in our experience such a condition can apply to another part of the body — the stomach. In some people, rodent ulcers can occur in it, causing (like tuberculosis) epileptic attacks.

The word rodent indicates something which is of a gnawing nature, and it can be applied to animals such as rats and squirrels — and to human beings, apparently.

Overheard conversation:

"Isn't it a shame about John Smith? He has been taken to hospital with TB. What is going to happen to his two girls? One of them has epilepsy; the other wears a uniform and goes to a high school of some sort."

Part 2

It is supposed by the scientific world that life began in the oceans about 1,000 million years ago. First there was DNA, the life molecules, which came together and formed amorphous life groups. By extremely slow degrees these evolved into creatures with light-perceiving organs (eyes) and, later, fins and tails. These were the fishes.

After a great period the fins developed into legs, and the creatures were able to breathe air as well as to exist in the sea. They could live on land and in the oceans. They were the amphibians.

As this was taking place, plants had evolved to grow on the land and to produce fruits and vegetables, which could be eaten by the primitive animals. By degrees, animals (some of them) found that as well as living on vegetable matter they could also live by eating other animals. Some of them did this exclusively. There became a division of animals on the earth: herbivores, living on plants, carnivores, living on animals, and omnivores, able to digest both sorts of aliments.

This was the situation about 200 million years ago on the land and also in the air, for by evolution some animals had developed their forelegs so that first they were able to glide and then, finally, to fly. These birds were divided into herbivores, carnivores and omnivores too.

200 million years ago conditions on the earth were excellent for life.

Due to the moon's proximity, the planet was warm with plenty of water, plants and animals.

According to ornithologists, there were plenty of primitive birds. All life in that remote epoch was primitive. All the animals and birds were cold-blooded and they ranged greatly in size. Compared with today, some of them were giants and some were quite small, but they all came under the name dinosaurs. They governed the land and water, and this was the situation for millions of years. Then, in a relatively short period, they all died out. They all, with the exception of three types, disappeared from the earth — the land dinosaurs, that is. This undoubtedly happened because, over the years, the moon had been receding and also animals had been evolving. At some period warm-blooded animals appeared and this, we think, was the main reason for the reptiles' disappearance. Another reason was that the earth became cooler so that the warm-blooded creatures had a big advantage over the cold-blooded ones. They could move faster and withstand the cold better, but it could be said the conditions towards the earth's equator (the tropics) are, and always have been, warm and equable. How could the dinosaurs die out there? But there it is known that the majority were very large and the warm-blooded (and probably amphibious) carnivores were comparatively small. These tended to attack in packs, and against these the giant dinosaur had little chance. These attackers were almost certainly the ancestors of wolves, and the saying is, even today, '. . . as a pack of hungry wolves'. Modern-day dogs can certainly swim — it is an ability handed down to them from years ago. If these occurrences took place on the northern and southern parts of the earth, there is little wonder that the dinosaurs tended to die out. What is left of them now includes the crocodile, found in the tropics. There are only two other kinds of creatures left in the world from the reptilian age. These are snakes and lizards, most of which, like the crocodile, are only found in the warmer parts of the earth.

All birds are warm-blooded. This great change from reptiles must have taken place before they learned to fly. Most nest in the tops of trees or in some other safe location — a situation which has greatly influenced their survival to the present time. They all seem to be descended from an animal similar to the pterodactyl, a long-extinct reptile. But many of them, when they were out of the nest and flying about in the daytime, took to roosting at night in trees, some of which were quite small and near streams or rivers.

This presumed fact leads us to discuss another animal which was very active at that time. This was a small cat-like creature which, as well as being able to catch fish with its claws, also developed a tendency to climb trees at night in order to search for roosting birds and large insects. These became its prey, and so successful was this creature in its way of life that some members of it took to residing permanently in the trees and living mostly on fruit, nuts, etc. With evolution going on all the time over great periods (millions of years), these small tree-dwellers evolved into monkeys, chimpanzees and, later still, into land animals that only slept in trees at night. These were large fighting animals resembling present-day orang-utans and gorillas.

As evolution continued, some of these ape-like creatures, instead of using their forearms as legs, began to stand upright on their hind legs and make their homes in caves, driving out the previous occupants (bears and the like). They learned how to make fires and lit them at night at the caves' entrances to frighten off wild animals, of which there were many. These were cavemen, who hunted by day and slept in the caves by night. They were hairy and very primitive in their appearance. They had small receding skulls, and communicated by a system of grunts. In the tropics, where there were many thick forests, their dark skins acted as a camouflage, and protected them both day and night. In northern regions, however, where there was sometimes snow and ice, these men were of a much lighter complexion. This afforded them protection also.

These men later learned to build habitations and to cook, etc., etc. This was what is now called the Old Stone Age. In Europe Neanderthal man was prominent, being replaced after a period by modern man.

So we have come a long way from the cat-like creature which existed many millions of years ago. Modern man is his direct descendant. But, strange to say, the cat-like animal is still with us. A modern domestic cat is very similar in every way to our ancestor. It likes to climb small trees (it will not climb tall ones) and it kills and eats roosting birds. It also likes to eat fish. When dogs descended from carnivorous wolves they would not eat fish. Cats are often active at night, when they catch small mammals, cats themselves being one of the first of these on the earth.

There has recently been put forward seriously by a newspaper that Homo sapiens is descended from an aardvark. This is an animal which lives normally underground. It has a snout like a pig, large ears and a thick tail. According to us, it does not resemble in any way monkeys, chimps (the most human-like of all animals), orang-utans or gorillas.

This theory seems to us to be ridiculous when we are told that every bone in a cat's body is present in that of a man. He has canine teeth, is omnivorous and likes to climb any object or structure.

CHAPTER 4

Part 1

The brain has been studied in great detail by many people in the past, but we have never heard or seen an account of the reason in it for the frontal lobes. It is supposed that they have something to do with a person's personality. We have thought a lot about them and our conclusion is as follows:

Everything in this world is divided into positive and negative forces and we find that these can be applied to the frontal lobes. One of them acts in such a way as to draw everything towards an animal, or in this case a human being, while the other acts in quite the opposite way — that is, pushing everything away from him — and generates a self-preservation effect. We believe that it is these two frontal lobes which play a fundamental part in the actions and reactions of a person to all his events — that is, everything he does.

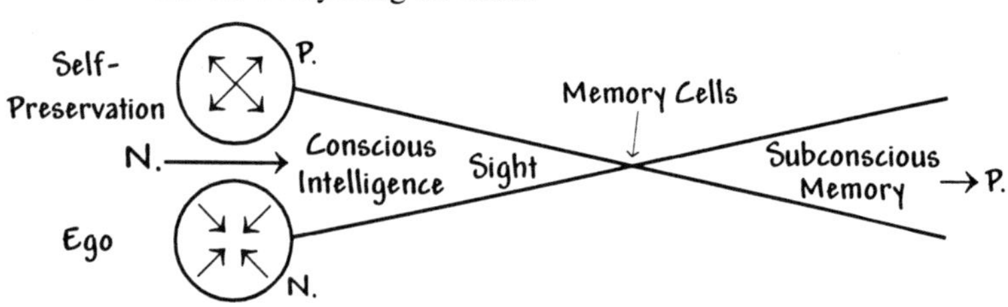

In this figure, which is supposed to represent the structure of the brain inside the skull, we can observe the frontal lobes and their positive and negative charges. If we take the negative lobe first, its forces have an inward direction, and it is this we think represents the person's ego. Putting this simply in a few words, this transpires:

Negative	Positive
1. I am God 2. I am perfect 3. Everything belongs to me	I must push everything away and be hostile to it

We can tell by our everyday experience how these two forces maintain a balance in the brain, and if there is an imbalance, what might occur. If the ego lobe is predominant, then the person has a religious tendency and is all for peace and goodwill towards men, he is self-assured and proud and has a good temper.

Unfortunately this lobe can have a bad side to it. We refer to theft. All thieves have the ego lobe abnormally developed. They tend to attract things or simply pull them towards themselves. Another undesirable effect is that of greed. Greedy people have a strong ego lobe, or it might be the effect of substandard intelligence, which plays the major part in their actions. For example, if a stupid person sees something desirable in a shop, he might be tempted to steal it and do so, whereas an intelligent person wouldn't.

Yet another example of the action of the negative lobe is a tendency to make a god of something or indeed somebody. All primitive men have gods of some sort. Belief in such boosts their ego. In civilised human beings, such a tendency is loosely called love. A person forms an unreasoning affection for another — this is called falling in love. It may not last very long, as the number of divorce cases in Western countries testifies.

As regards this, it has even been reported in America that monogamous couples are almost unnatural and that the primitive men from whom they are descended were polygamous. There seems to be something in this notion when we observe that married couples frequently quarrel and part, whereas in polygamous associations women get along very well with each other.

But what about the other lobe — the positive one? What effects and actions does it have on humans? Just as the negative lobe generates love and attraction, the positive one generates the opposite of these — namely, hate and animosity, bad temper and aggression, wars and dislike. Boxers have this lobe very strongly developed, as do men wanting to join the army or one of the other fighting services. In another context, though, the positive lobe can generate fear — the person becomes afraid of something. This can occur as a disease, he becoming depressed.

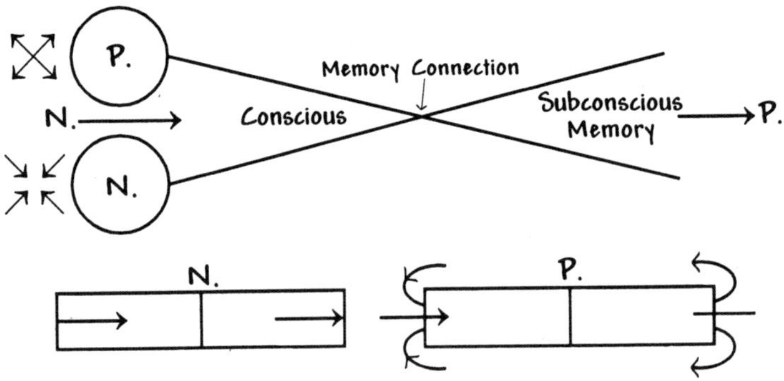

These figures represent the difference between positively charged iron and that which is negatively charged. It is obvious to us that the brain resembles the former, so the brain must consist of iron in some amount. It is divided into conscious and subconscious parts. The centre, as in all magnets, has the least strength; the strongest is at the ends.

As we all know, the brain contains the senses by which animals are aware of their surroundings. As well as this, these senses are connected to the memory, where the information is stored. We don't know what the memory consists of, but we suspect it has something to do with the 10,000 million cells which make up the brain. We can only imagine that electrons form brain cells into patterns of charges throughout life, forgetting nothing. We all know also the weakest parts of magnets (positive ones) lie in their centres and we think that this in animals divides their brains into conscious and subconscious parts. If the connection in these is weak we imagine then that the animal, including man, has a poor memory and consequently if it is strong he has a good one.

Some people, we know, have total recall of their memories, which only means that their connecting tissues are very strong.

Why, we might ask, if the brains of animals are basically magnetic, do we not perceive this externally? The simple answer is that the brain's magnetism is extremely weak compared to positive magnets, and its inner workings cannot be perceived at all.

But there are some people who do possess the ability to perceive some of these — notably, gipsies. Some of these, as well as having the ability to divine thoughts in other people, also profess to be able to divine their future, and some of their predictions are remarkably accurate.

Why should this be so? We can only hazard a guess ourselves.

Life can be compared to a gramophone record: its present is what the needle is actually playing; the past is what has been played; and the future is what it is going to play. It is the future which is the mystery. This must be like the memory patterns of magnetic charges, and they seem to be placed in a human's brainstem, for cutting this off or damaging it in some way results in instant death. Fantastic? It appears to be a sober fact.

We might come now to another apparently mysterious fact or occurrence. This concerns hypnosis — the power which some people possess of making others unconscious. This fact, like some others, has a simple explanation — simply that the positive and negative charges in the brain have been reversed. The conscious becomes the subconscious, and the subconscious becomes the conscious while the subject is unconscious. During this time the hypnotised person has total recall and can, if questioned by the hypnotist, give detailed accounts of things that happened to him many years before, which he could not attempt to do in his normal state. It should be explained here that positive forces, like those of gravitation, are illimitable and can be applied to those of the memory.

We can make a note here and can apply it to a Dutchman. He fell off a ladder one day and was carried off unconscious. When he woke up in hospital he found that other peoples's thoughts started to occur to him. We can explain this by looking again at our figures. It is obvious from these that the Dutchman must have fallen on his head — the front part of it — so that it became abnormally sensitive, and he could pick up other people's positive memories. We do not know what subsequently became of him, but his case seems to go a long way to proving our theories.

Part 2

Much has been written and discussed about the colours and origins of man, but none as far as we are aware have pinpointed his origin as being the East Indies. There, as described in a previous text, apemen going on all fours learned by swimming and being buoyed up by water in those water-girt lands to walk upright. His subsequent superiority over his ape rivals enabled him to increase and spread all over the world. The progress and extent of the migrations can easily be followed. First, going north by sea in crude boats and by land up the Malay Peninsula, China was invaded. Then, travelling west through Burma they reached India. After this, going north, they invaded Russia and, at the same time, to the south and west, they overran Arab countries. It was only a step from these to North Africa and subsequently to the south of that vast continent.

While this was taking place, Europe was populated. Of a certainty these migrations took many years and many wars were fought backwards and forwards over disputed territory, just as is going on nowadays. Going east the primitive men from the East Indies had, at the same time as the other invasions were taking place, spread across the Bering Strait into North America and then south into those vast lands past Panama.

As his geographical situations needed it, man took his coloration. The original colour of his skin and hair — for he was hairy — was dark, for he had originated, it must not be forgotten, in the dark recesses of jungleland, where pale coloration would be fatal. Different colorations are now to be seen all over the world, from the desert yellow of the Asian-race Chinese to the black jungle-inspired Negroes, thence to the dark-brown Indians and the red counterparts of the North and South Americas. These Americans obtained their colour from the yearly fall of red leaves, these making a red skin a good camouflage in frequent contests with the life-producing buffalo. But what about Europeans? In Europe the most dangerous time of year was when snow was on the ground, hence their white complexions. This observation particularly applies to the North European races.

So now how did modern man descend from East Indian savages? We can't be certain of these things, but a couple stand out. The first is what is known as atavism — that is, the appearance of ancestral (not parental) characteristics in human beings. These take the form of dark patches, often raised, which occur on humans all over the world. These are full of dark, sometimes black, pigment. Also, sometimes these dark patches are covered with hair — occasionally only a few hairs — suggesting that man had a dark-skinned ancestor who was hairy. Both males and females of humans can have such patches, and the fact that they are universal in black and white races shows conclusively that they had a common origin.

The second thing that points to the Indies as being the cradle of humans is the theory — and it is only a theory — concerning the monthly periods of female humans. When we say monthly, we immediately think of the moon. What occurs monthly on the earth to do with that satellite? Why, the tides, of course. Can we relate the monthly menstrual flow of female humans with water — the tides? It has been described before how probably in the East Indies humans, by being in the sea (water) gradually learned how to stand and walk upright. Then he practically lived in water. He learned also how to swim, and he made crude boats to help him fish. It is known that nearly all mammals have a menstrual flow and that these vary in time according to the circumstances and lifestyle of the animal.

Many years ago, humans in the East Indies had a lifestyle which was closely associated with the sea and the tides. Cleanliness, as is usual in humans, was at a premium and those who lived there in those tropical islands were cleanly animals. They were aware that flood tides were at a peak every month and that then the high-water mark came much higher than normal. They knew the extent of these monthly high tides and so arranged that their middens were situated where every month the flood tides would come in and clean out the communal refuse. How were the females affected by the tides? Since they were often in the water, the extent of it affected their menstrual flow, which came to coincide with the high monthly tides. This was a cleanly process and the females were more healthy as a result. Whole families were involved in these events and they continued for many years until in fact man evolved further and left the seashore. After this man covered the globe, and though his females still maintained monthly menstrual flows, their times were all mixed and differed from each other. These things are only theoretical, however, but it is nevertheless a remarkable coincidence that a woman's periods occur at monthly intervals.

So, from the facts of dark patches occurring on humans, together with a female's monthly periods, we can deduce that man originated from dark hairy people who lived in tropical islands where they were very closely associated with the sea. Such conditions we might also associate with the West Indies, of course, but there there are no apes and no ancient human remains such as are found in the East Indies. Even today, Pacific Islanders have a great affiliation with the sea and water and will display considerable prowess in retrieving valuables thrown into the sea from ships.

We are aware that the modern view is that man originated in Africa. We might dispute this because of the foregoing. It is also stated that man's loss of body hair has never been explained. Darwin's theory was that a change of that kind is caused by sexual selection by males who preferred their mates to have furless bodies, the effect of this selection being partially transferred to men. Animals exist with thick fur coats in warm climates, so why not man? We have only to look at animals similar to man in the tropics to obtain a partial explanation of this. The gorilla, chimpanzee and orang-utan are one and all tremendously strong and are great athletes, whereas humans are puny in comparison. Man makes up for his lack of strength, partially by his intelligence, and also by, we suspect, his lack of hair (fur) on his body. There is nothing like hair for retaining scent and this in jungle and other conditions would be a great disadvantage to man as many of his enemies (including big cats) have an acute sense of smell.

The fact is, we think, that Darwin's theory is partially correct. Males may have chosen furless mates, but what is much more likely is that women's lack of hair is part of their defence mechanism — they are less likely to put off strong scents than their male counterparts. They are the weaker sex and are passive when it comes to protection. But there is another side to the absence of hair or fur on humans. This concerns clothing. Many years ago, probably during the last ice age, northern man started to wear clothing. He found out that by wearing the skins of animals he killed he was warmer and had better protection from injury. He also began using leather on his feet for the same reason. These advances spread from tribe to tribe until, after a certain period, almost all humans in cold climates wore clothing. What was the result of this advance in man? It is well known that animals', including man's, parts or characteristics tend to disappear if they are not used. What then happened to man's hairy or furry body when he began wearing clothes? Why, the hair or fur became less — man exchanged a natural furry coat for a skin one. Later this change became more pronounced when he covered himself with wool. Today man's fur has almost all gone; that which remains is at the joints of the limbs with the trunk. He still has a head of hair, of course. Hair on females has a sexual significance — males are attracted by it.

It is nowadays usual to attribute man's colour to sunlight and the amount of ultraviolet rays that are contained in it. We are told that, in order to protect himself from the latter, man in the tropics has evolved a dark or black skin, which ultraviolet light cannot penetrate. We are further told that white races have a light skin in order to allow beneficial rays from the sun to penetrate it. It is our contention that these reasons for colours of men are only partly true. Consider men in the dark jungles of the world. The sun's rays cannot penetrate there so we would imagine that men living there would have pale skins so as to allow what little light rays — beneficial rays — there were to penetrate them. But what do we find? Universally people inhabiting them are dark-skinned or black. This seems to suggest what we suspect, that such a coloration serves as camouflage in the presence of danger. In the case of the white-skinned races it might be thought that such people living in northern lands would be dark-skinned in order for them to absorb the health-giving sunlight. So it does seem that, in part, their colour serves as a protection against dangerous beasts. There are not many of these now, but during the ice age they were numerous. However, if we cross the Atlantic and journey north to the Arctic regions we come to a race of men who are certainly dark-skinned. These are Eskimos. Despite the conditions in which they live they are dark-complexioned, showing that our reasoning as to coloration might be correct.

There is another part of the globe which we have not touched on yet in this discourse — this is Australia. The original inhabitants there (the aboriginals) were called by sailors who first encountered them 'the most miserable wretches on the face of the earth'. This referred, no doubt, to their having no husbandry and to their willingness to eat almost anything. In addition to this they had no dwellings but lived in humpies, which were made of strips of bark slung over two sticks. They were naked and black, a colour which they had doubtless obtained by living in the Australian outback. They had spread all over the Australian continent. They were armed with spears and were — as the North American Indians were when they met the white men — almost defenceless. Even then they had a curious weapon for disabling animals — a boomerang. This was made of wood and when thrown it tended to return to the thrower. What dangers were there in the outback to produce their colour? Probably more than nowadays. The dangers would include snakes and crocodiles. Where did the aborigines originate? Probably New Guinea, though many of the inhabitants of that island are of Negro origin.

CHAPTER 5

Part 1

There is no doubt that man has been a fighting animal ever since he came down from the trees. It is also quite probable that he was polygamous — that is, he had more than one wife. This is still the case in many Arab and African countries. A man can have as many wives as he can afford. This, we think, is a very good practice since it leads to the best men having the most wives and children — these latter tend to have the same brains as their parents.

This is not, of course, the case in the West, where monogamy is practised. There a man can be sent to prison for possessing more than one wife. Recently, in the United States, a man was jailed for possessing eight or nine wives. How he managed to keep them was not made clear, but it was noticed that they all got on well with one another. This is not often the case in Western countries where married couples often quarrel and part. A lawyer investigating scores of divorce cases came to the conclusion that he doubted whether monogamy was natural in human beings. It used to be said that a sailor had a wife in every port, and this observation was in many cases quite a fact. A man can father dozens of children, whereas women can only manage to have a few. In the West a lot of children are unwanted.

It was stated by a well-known actress in her autobiography that she did not know why so many white women are attracted to coloured men, and it seems to be rare that these have difficulty in finding a wife. Going back in time to the mutiny on the *Bounty*, it was noticeable that its sailors had no difficulty in finding young native women to accompany them to Pitcairn Island, where some of their descendants live today.

It also turns out that hundreds of coloured women in East Africa had children by white British soldiers and sailors.

It also seems to be a fact that very few women will marry a man they dislike, even if it means that they will be left on the shelf. In the East, however, it appears that women are glad to be chosen with others to be the wives of rich men and to be included in their harems. This means that they will be well kept and have children, which most want.

It was stated recently that a very well-known figure in the Arab world was the forty-seventh child of a rich furniture manufacturer. Of course, all the children had plenty of money and had little need to work, but most of them did.

In view of all this we can state that in the world of the apes it was noticed that the females of one tribe were most attracted to, and wished to mate with, males of a quite different one. This seems to be instinctive in them, and that is a defence against too much inbreeding in the tribes and shows a desire for fresh blood to be maintained in them. In this way the population is protected against diseases which tend to increase in a tribe which has little or no defence against them. The addition of fresh blood limits this.

We might conclude this account by considering what men and women in the West prefer in individuals of the opposite sex. In men it seems to be simple: they most like women with good looks and personality or just personality. Women like strong men whatever their looks, and they also have a preference for foreign men or even men with just a foreign accent, in the same way as female apes prefer to mate with male apes from a different tribe.

Part 2

Since the beginning of time man has always had a belief in God or gods. By believing this, he associated himself with that God or gods, and this boosted his ego. This ego is almighty, of course. From the beginning of history, leaders of men have had their egos boosted and this has had the effect of empires being created and dominions established. Primitive men — and there are millions of them about, even today — are willing to sacrifice themselves for their religions in the belief that they will go to an imagined heaven of some kind.

This is certainly not the case with most civilised men, which fact leads us to the inevitable conclusion that all men are not equal and that Darwin's theory of the survival of the fittest is certainly correct.

We think it is deplorable in this day and age that the action of the basic frontal lobes of man, which give him his character, are not fully understood by physicians or doctors of medicine. True, there was a time long ago now when operations were carried out on mental patients to sever their lobes' connecting fibres from the other parts of the brain, but these are not now carried out. The reason for this is that if both of the frontal lobes' connections to the other parts of the brain are severed, a cabbage mentality of the person involved results. The reason why this happened was that the medical men did not know which lobe was faulty. The lobes did not respond to any electrical stimulus. They were called the 'silent' areas of the brain. If the correct lobe's joining fibres were cut we think the patient would be cured.

We heard a story of two brothers. One was a high-up in the religious world; the other was an habitual thief. The latter had a frontal lobotomy, and afterwards he found that he had lost his thieving habit.

There is a great tendency nowadays for many members of the population to lose their faith in religion — they cease to believe in God. We think this has had the effect of the positive lobes of their brains becoming predominant. Such people become more and more aggressive and violent. Muggings take place, more murders are committed, churches and burial places are vandalised and all sorts of violent crimes are perpetrated.

In regard to the fact that more wars have been caused by religion than anything else, we can only say that the ego lobe has been depressed in some way, leading up to the aggressive becoming very strong. The result is that conflicts ensue. The crusades are an example of this. The ambition was to throw out the Arabs from the Christian shrines. They failed, but the present-day Arabs still regard Jesus of Nazareth as a minor prophet.

The Romans had gods of some sort, of course. One of the principal ones was Mithras. When the majority of heathen and other parts of the Roman Empire became Christian and ceased to worship that god, Mithras' Day (25th December) became the supposed day of the Christian God's birth.

Many things have been added to that belief: Advent, the Three Wise Men, the birth in a manger, the Virgin Birth, etc. Looking at these things, it seems to us that there is no wonder that many modern men have no religious belief and that crime is rampant.

Primitive men still have their beliefs and animosities. An American's recollection of races and religions some eighty years ago was that the city in which he was born was divided in this way: the blacks hated the whites, the whites hated the blacks, the Catholics hated the Protestants, the Protestants hated the Catholics, and they all hated the Jews.

Jews, we all know, are among the cleverest of men. Wherever they have gone since the Romans forced them from Palestine, they have made their mark. This fact, as well as their not being Christians, has, we think, always led to their being disliked and often persecuted.

The average Jew, it would appear, is more clever than the average European or North American, and we only have to look at the Christians' Bible to find the reason for this. There, the Jews' history is one of war — perpetual war. According to Darwin's theory, this could only have one result — that the fittest would survive. In the Jews' case this means the most clever.

It seems, looking at Europe, that it is the most primitive peoples who are the most deeply religious. We are, of course, referring to the Balkans and Ireland. In both these cases religious hatred is predominant, and the most cruel things are perpetrated. Why should this be so? The answer is that the ego of passionately religious people is so strong that they are easily offended, the strength of the ego is sharply reduced and the positive lobe becomes predominant. This makes them commit acts of violence in order to right the ego.

This is certainly true in the Balkans, whose inhabitants can be called peasants, and it is certainly true of similar people in Ireland. There, many Irish seem to have the same mentality as peasants of long ago when most of them lived principally on potatoes, went barefoot and lived in hovels. Their religion is based on Jesus of Nazareth, who is their God. 'Holy Mary, Mother of God' is one of their most frequent expressions, and this sort of religion prevents them interfusing properly with the other countries of the British Isles.

CHAPTER 6

The very fact that humans — Homo sapiens — whatever their race or colour, can interbreed points conclusively to the fact that they are all descended from the same race of man.

Also, in relation to this we must be aware that that particular race of people who lived close to the sea found it very difficult to obtain pure water. Man could not drink seawater and had to rely on river water, which was never pure, for his sustenance. He did not have the knowledge or means to purify it, as modern man has.

What effect has this had on his development over very many years? Well, we know that it is the liver of men which is most affected by what they drink, and it is this organ that suffers most from the harmful impact of impure water. The impurities had the effect, we think, of changing his liver so that it can repair damage to it sustained by impure water or other deleterious substances.

Physicians are aware that the liver of man has the power, if damaged in any way, of regenerating itself and growing — even if a part is destroyed — back to its former size.

This is an ability that no other organ has, and it has developed simply as a necessary protection against impure water or other liquids that man may imbibe.

However, it cannot defend itself absolutely from diseases such as hepatitis or alcoholism, which sometimes cause death.

CHAPTER 7

Part 1

Turning to entomology, we can consider the ordinary housefly. It has black and grey stripes on the back of its thorax, and its abdomen is of a dark yellow colour. It is hairy, the hairs being very large compared to its size. It probably uses these hairs as organs of touch, enabling it also to judge its speed and position in flight. It has been reported that its ears are in its front legs. Its sense of smell is remarkable. It only perceives decaying matter, for while it remains unmoved by the scents of flowers, it can perceive very strong odours at a distance. Strange to say, the housefly itself has no perceptible odour. These things are fairly common knowledge, but not so commonly known is the fact that, almost unique in the insect world, the housefly has no alimentary system ending in an anal passage. It is true that it eats food, which must pass to a stomach, but there the similarity with other insects ends. Instead of its digested food then going to intestines and then being voided, the housefly then throws it up out of the suction tube which serves as a mouth. The rejected substance is left wherever the fly is settled at the time. The marks left by houseflies on walls and other places, such as lampshades, are often assumed to be fly excrement, but this is not so. The fly abdomen is used as a developing place for eggs in the female, and for sperm in the male. There is no anus or vent in houseflies.

Such insects must have developed where there was — to put it mildly — plenty of refuse and dirt, and in countries where there was a warm climate. They all die in the winter in northern countries, leaving their eggs to develop in the next summer.

It has been estimated that they have been almost unchanged on the earth for at least 60 million years, so that it seems likely that their main predators (spiders) have been around for a similar period.

In all our lives of experience concerning the oceans and seas we have never heard a single word in favour of sharks. They are always portrayed as ruthless killers which have no role to play in the marine world, and which should be got rid of whenever they appear. They are considered dangerous predators of anything live in the sea, including any human beings who might venture into it.

But we think nothing is further from the truth. Sharks play a very important part in the ecology of the oceans. Fishes live and die in them, and whoever heard of dead fish being caught up in the nets of fishing vessels? The fact is that sharks have a keen sense of oceanic odour and very sharp eyesight. They can also detect when other inhabitants of these oceans are failing in health and are likely to die.

Sharks are attracted to such creatures; the sharks kill and devour them.

It is in this way, we think, that the seas and oceans of the world are preserved in a cleanly state without rotting fish polluting them.

This simple fact also applies to animals on the land, the survival of the fittest always being an indisputable law there.

It is also a fact that if the blood of several human beings, minus the red corpuscles, was placed in a tank, tropical fish could live in it. This shows, without any doubt, that life first began on the earth in its oceans.

Part 2

What would be the consequence on the earth if its surface heat was increased by — as it almost certainly will be — a further five and probably more degrees?

First we might imagine that a consequent rise in temperature in the earth's interior would result in an increase of volcanic activity and earthquakes. Of course, these phenomena go on all the time, but also it is said that the polar ice caps will melt with a subsequent rise in the water levels all over the world. Many low-lying places on it will be flooded. These events must have taken place before, of course, and polar animals that survived made their homes on the many northern islands, such as Bear Island and Spitsbergen.

There has been an enormous amount of talk recently about poverty in Africa, and the shocking fact that on average 4,000 children there die every day. Much money has been poured out and vast concerts held in support of the Africans.

This planet, it is known, is terribly overpopulated. The present situation takes my memory back to the last war when there was a very popular double act on the stage and on the radio called the Western Brothers. One of these played the piano while the other recited monologues and told jokes. One of these sticks in my memory. It went like this:

The first brother said, "In regard to the crocodiles in the Nile, during their mating season the female lays ten eggs and the male comes along and eats nine of them."

"Well, what about it?" asked the other brother.

"Don't you see," the first one said, "if this didn't happen you would be up to your neck in crocodiles."

This story applies in this day absolutely to the poor Africans, who keep having children at a tremendous rate, with nothing to stop it. It's very obvious today that what the African natives want more than anything is contraceptives.

We think, in regard to this matter, that the Chinese have the right idea. There you can be put in prison for having more than two children.

Instead of pouring money into Africa — a lot of it inevitabley being wasted — we think some of it should be spent on contraceptives, despite the fact that fertility is greatly admired by the natives.

We say, as we have said before, that as the world is overpopulated, we should think first of reducing the peoples in it, and that contraceptives are the only answer.

CHAPTER 8

Much has been said and written recently about metal being weakened so as to break under the influence of human hands. We might now try to explain the reasons for this. All animals and plants are made of cells. These are very tiny but are basically all the same. They consist of a nucleus which is surrounded by a jelly-like substance called the cytoplasm. This is transparent and, like the nucleus, is composed of DNA molecules. Both the nucleus and cytoplasm are enclosed by a very thin membrane, called a plasma membrane.

From our way of looking at it — suspecting to find evidence of positive and negative influences — we are not surprised to find these present. The nucleus, we are informed, is negatively charged and the cytoplasm and plasma membrane are positive in their nature. What these words mean really we can only guess. It seems feasible that the nucleus is negative because of its density. This cell brain, as it is called, contains all the genes and chromosomes which will subsequently build up animals and plants. The density of these nuclei must be extreme and can be compared to great masses. These, when slowed in their movements, give off electrons and so we can expect that the cell brain does this too without moving, giving off minute electric currents, both for directing the working of the cell and building up very complex organisms, utilising the probable magnetic charges of the genes for this purpose.

The cytoplasm, as stated, is transparent and this almost certainly indicates that it is composed of positively charged matter. This is the giant life molecule, called DNA, which makes the difference between living things and inanimate ones. The living ones are magnetic and very complex spirals which can divide and therefore increase in size, while inanimate molecules cannot. Although all the atoms in DNA molecules are combined with each other, and so use up a great amount of their external forces in combining, the general effect is positive. Their bared charges render DNA molecules acidic in their reactions with other atomic groups and also with each other. Indeed, DNA is called DNA acid by biologists. The A stands for that quality. It is the property of acids to have what we can call a gnawing effect on matter with which they come into contact. They do this simply by being very positively charged — their twin field forces are bared and therefore are attractive to electrons. These carry a negative charge and respond to acids by leaving their orbits and thus weaken any substance of which they are a part. It should be understood here that electrons hold matter together, utilising positive and negative forces, the positive causing movement which takes place in the direction of the negative attraction. In atoms these forces are very strong and the electrons can be called linchpins, holding atoms and therefore molecules to each other.

We must now consider the cells as a whole.

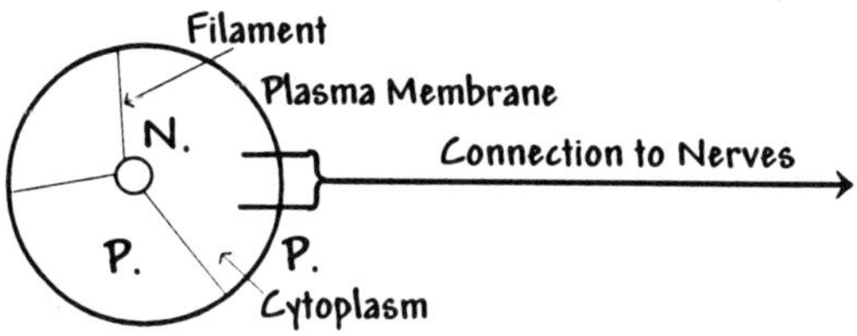

They bear a resemblance to a hen's egg — the nucleus, the yolk; the white, the cytoplasm; and the shell, the plasma membrane. But we must here use again our imagination. So little is known about the general action of cells we are forced to do this. We have explained the probable structure of cells — that is, the cell brain and the cytoplasm composed of positively charged DNA molecules. The latter hold a certain charge and are connected in organisms to their nervous systems. We now venture on the unknown. We can suppose that the plasma membrane is positively charged by receiving a constant electric current. If this were so and it came from the nucleus, then this would serve to hold the cytoplasm charges in and prevent at the same time atoms from outside entering the cell, much as the shells of eggs serve to protect the white and yolk and keep other matter out. But atoms from outside cells constitute what we can call food and so the nuclei of cells can, by limiting or increasing their electric currents to their plasma membranes, control the entry of food to their cells. Food means also growth, and so it can be seen that the growth of cells — always by division — in animals would always depend on the strength of the negative charge of their nuclei.

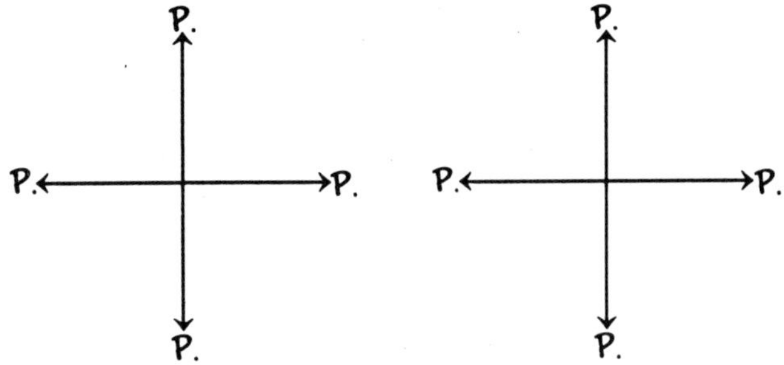

Animals eat food and this contains all the atoms that cells need to promote division and growth. We can see also that all animals grow to a certain size, live their lives and then die. This is what happens to their cells, but sometimes this process goes wrong and benign and malignant growths occur in organisms. It is obvious that these could be caused by the negative nuclei being weak and not preventing the admission of food atoms to the cells. Such cells would have unlimited growth therefore, and this would probably have a very serious effect on the organisms concerned, leading in many cases to death. Simply, if the above were the case, then it would explain all growths in animals, including fishes, by stating that if the nuclei of their cells were weak and put off weak electric currents to their skins (plasma membranes), then the positive charges of these would be weak too, and there would possibly be abnormal growths.

It follows from this that as it is very likely that all the charges in organisms become weakened with age, the chances of abnormal growths in them become greater, and so the older they are the more likely they are to develop benign and malignant growths. It follows also that the cells in growing animals must put off weaker currents to their plasma membranes in order to facilitate growth, and that sometimes this process goes wrong and young animals develop tumours. It is also possible that the process of maturing in animals, when their various parts are developing, notably sexual, causes the nuclei of the cells involved to put off weaker currents to allow for this and this going wrong causes tumours to occur.

We have with all this got off the subject of why some subjects have the ability by touching them to cause metals to break up. We know now, or think we know, that all atoms are positively or negatively charged, their cells being positive. These are connected to the central nervous system, constituting the animal's sense of touch. We know also that it must be the tissues of humans — those of the fingers — coming into contact with metals which cause these to come apart when touched. It must also be a fact that some humans' positive magnetism is abnormally strong. This appears in their hands and therefore their fingers, so that when these rub against metals the electrons holding these latter together, being negatively charged, absorb this positive force and are thrown out of the metals, thus weakening their structure. If the finger rubbing then continues, the point may be reached when so many electrons leave the metals that these break up. We know that metals are all positively charged, meaning that they have few electrons. These electrons cling to the metal's surface constituting a sort of glue so it is not so difficult, using a positive influence, to cause them to fall apart. It would be almost impossible to affect this with other matter, where the electrons are situated internally.

The fact of the moon having a great effect on the earth's climate has had, apart from its hot and cold cycles, some remarkable results. Not least of these has been the moving of millions of tons of earth and rocks from one place to another. This happened in Europe during the last ice age, which came to an end about 10,000 years ago. During that ice age great glaciers were formed, moving from high to low ground. These scraped over the earth very slowly, displacing in their progress huge amounts of matter. The consequence of this was that when finally temperatures rose and the glaciers melted they left behind all the matter they had displaced. A good deal of this was composed of tiny particles, which subsequently made earth and pasture. This was utilised by man when he learned to farm and keep animals.

All this being said, we must look more closely at what happened to plants during the ice age. Most plants will not grow when the weather is very cold. Their seeds remain dormant in the ground. In Europe these seeds were covered by glaciers. These carried earth, and, upon their melting, seeds which had been on the surface of the ground were now buried several feet deep — so deep, in fact, that they remained dormant and did not grow when warmer weather supervened above them.

Now we must go back a little in time to 1914 and the war that started then. As is well known, the opposing armies became bogged down in what had been rich farmland and, as a consequence of this, began to shell each other without mercy. If these shells missed their targets, as a great many did, they exploded harmlessly in the ground. But inevitably it happened that this was just what was necessary to force seeds previously buried in the ground to come to the surface, and many of them were able to germinate there in the warm weather. So it is apparently possible for seeds of some plants to live for thousands of years in the earth and then, if conditions are right, to germinate. It was with some surprise in the First World War that soldiers saw poppies growing all about them and in shell holes. It seems that this plant is especially hardy, but it may have been noticed primarily because of its colour, which is bright red. Most grasses growing in the same conditions went almost unnoticed.

We should remember these facts when the next Poppy Day comes round.